Russia's Securitization of (

This book provides an in-depth analysis of how mobilization and legitimation for war are made possible, with a focus on Russia's conflict with Chechnya.

Through which processes do leaders and their publics come to define and accept certain conflicts as difficult to engage in, and others as logical, even necessary? Drawing on a detailed study of changes in Russia's approach to Chechnya, this book argues that 're-phrasing' Chechnya as a terrorist threat in 1999 was essential to making the use of violence acceptable to the Russian public. The book refutes popular explanations that see Russian war-making as determined and grounded in a sole, authoritarian leader. Close study of the statements and texts of Duma representatives, experts and journalists before and during the war demonstrates how the Second Chechen War was made a 'legitimate' undertaking through the efforts of many. A post-structuralist reinterpretation of securitization theory guides and structures the book, with discourse theory and method employed as a means to uncover the social processes that make war acceptable. More generally, the book provides a framework for understanding the broad social processes that underpin legitimized war-making.

This book will be of much interest to students of Russian politics, critical terrorism studies, security studies and international relations.

Julie Wilhelmsen is Senior Research Fellow at the Norwegian Institute of International Affairs (NUPI), Norway and has a PhD in political science from the University of Oslo. She is co-editor of *Russia's Encounter with Globalization: Actors, Processes and Critical Moments* (2011).

Routledge Critical Terrorism Studies

Series Editor: Richard Jackson
University of Otago, New Zealand

This book series will publish rigorous and innovative studies on all aspects of terrorism, counter-terrorism and state terror. It seeks to advance a new generation of thinking on traditional subjects and investigate topics frequently overlooked in orthodox accounts of terrorism. Books in this series will typically adopt approaches informed by critical-normative theory, post-positivist methodologies and non-Western perspectives, as well as rigorous and reflective orthodox terrorism studies.

Russia's Securitization of Chechnya

How war became acceptable

Julie Wilhelmsen

Routledge
Taylor & Francis Group

LONDON AND NEW YORK

First published 2017
by Routledge
2 Park Square, Milton Park, Abingdon, Oxon OX14 4RN

and by Routledge
711 Third Avenue, New York, NY 10017

First issued in paperback 2018

Routledge is an imprint of the Taylor & Francis Group, an informa business

British Library Cataloguing-in-Publication Data
A catalogue record for this book is available from the British Library

Library of Congress Cataloging-in-Publication Data
Names: Wilhelmsen, Julie, 1969– author.
Title: Russia's securitization of Chechnya : how war became acceptable /
Julie Wilhelmsen.
Description: New York, NY : Routledge, 2016. | Series: Routledge critical
terrorism studies | Includes bibliographical references and index.
Identifiers: LCCN 2016017948| ISBN 9781138187139 (hardback) |
ISBN 9781315643359 (ebook)
Subjects: LCSH: Chechnëiia (Russia)–History–Civil War, 1994–
Public opinion. | Public opinion–Russia (Federation) |
Terrorism–Prevention–Government policy–Russia (Federation) | National
security–Russia (Federation) | Russia (Federation)–Relations–Russia
(Federation)–Chechnëiia. | Chechnëiia (Russia)–Relations–Russia
(Federation)
Classification: LCC DK511.C37 W54 2016 | DDC 947.086–dc23
LC record available at https://lccn.loc.gov/2016017948

ISBN 13: 978-1-138-54986-9 (pbk)
ISBN 13: 978-1-138-18713-9 (hbk)

Typeset in Times New Roman
by Wearset Ltd, Boldon, Tyne and Wear

Contents

Acknowledgements

There are many people I would like to thank for their support in helping me to complete this book, which is based on my PhD. First of all, I would like to thank Richard Wyn Jones, who has been frank, enthusiastic and supportive beyond what could be expected from any supervisor. I have benefited greatly from the environment at the Norwegian Institute of International Affairs (NUPI), and in particular from my colleagues in the Russia research group. Cooperation and conversations over the years with Helge Blakkisrud, Geir Flikke, Jakub Godzimirski, Heidi Kjærnet, Elana Wilson Rowe and Indra Øverland have enabled this book. Also the group of NUPI colleagues who persistently work on theory, despite the lack of funding for such activity, has been a great inspiration. I am particularly thankful to Kristin Haugevik, who really has been my closest companion on this journey, intellectually and personally. Several persons have read my text or parts of it at various stages. I would like to thank Patrick Jackson for very useful comments on my theory chapter and Jeff Checkel for thorough feedback on my research design. I am indebted to Elana Wilson Rowe, Iver Neumann, Ole Jacob Sending and Stacie Goddard in particular, for having read the entire dissertation and providing comments that were both encouraging and demanding.

On work in turning the dissertation into a book, I am thankful to two anonymous referees working for Routledge for very encouraging and helpful comments. Parts of Chapter 2 have been adapted from Wilhelmsen, J. 2016. 'How does war become a legitimate undertaking? Re-engaging the post-structuralist foundation of securitization theory'. *Cooperation and Conflict*. Published online before print, May 13, 2016, doi: 10.1177/0010836716648725.

Although the content in this book is fully my own responsibility, I have relied on several people for assistance with technical and linguistic challenges along the way. I would like to thank Natalia Moen-Larsen and Alesia Prachakova for helping me search for texts and Zhanna Petrukovich for helping me with legal documents and checking Russian transcriptions. I am thankful to the librarians at NUPI, Tore Gustavsson in particular. Last but not least, I am grateful to Susan Høivik and Wrenn Yennie Lindgren for their invaluable language editing. A special word of thanks to Jan-Morten Torrisen, Head of Administration at NUPI, for encouraging me while I was ill and creating sufficient space to enable me to complete this book. I would also like to thank the Norwegian Ministry of Defence and the Norwegian Research Council for generously financing this project.

Beyond academia, my family and friends have provided invaluable and continuous support. In particular I thank my five children Georg, Isak, Karl, Maria and Agnes for being so patient, loving and encouraging and reminding me every day what the most important things in life are. Above all, I am indebted to my dear husband Frantz, for everything.

1 Introduction

Russians were reluctantly dragged into the first post-Soviet war against Chechnya in 1994. By contrast, the Second Chechen War was launched with a collective call for violent attack. Charles W. Blandy argues that the main difference between the two Russo-Chechen conflicts is not in terms of military strategy, but in terms of the 'resolute firmness of the political authorities in prosecuting the war in Chechnya, having secured the backing of Russian society as a whole'.[1] Most Russians had considered a new war against Chechnya totally unacceptable only half a year before Russian ground troops again entered Chechen territory in the so-called counter-terrorist campaign in October 1999. However, when October came, hardly a voice was raised in protest against the massive violence launched against this Russian republic.[2] How was this shift made possible? In more general terms: how does war become acceptable?

Scholars agree that the brutality and the extent of war crimes committed during the Second Chechen War were as massive as during the First Chechen War. While identification with Chechen suffering inflicted by war increasingly constituted a pressure to end the First Chechen War, no such pressures emerged in Russia during the Second Chechen War.[3] How can acceptance of massive violence against fellow citizens continue, when the human cost of war is revealed? This book seeks answers to these questions by exploring the 'securitization' of Chechnya in Russia from 1999 to 2001.[4] Advancing a post-structuralist reinterpretation of securitization theory, it argues that representations of the Chechen issue in Russia during 1999 comprised a re-drawing of the boundaries between 'Chechnya' and 'Russia' in Russian discourse that served to legitimize the violent practices employed against Chechnya and Chechens during the Second Chechen War.[5]

This is not a study of *why* the Second Chechen War was launched and what the motivations were, but about *how* it became seen as a legitimate undertaking. The new military campaign against Chechnya was allegedly planned well in advance, but this book will not delve into what the Russian leadership wanted to achieve by it.[6] The focus is rather on how broad public acceptance for a new war came about in the first place, and how such broad acceptance was sustained as the war unfolded in all its brutality. This book reveals how the intensive, observable linguistic practices that served to represent 'Chechnya' and

'Chechens' as an existential terrorist threat to an innocent and victorious 'Russia' made violent practices such as those used in the Second Chechen War possible and acceptable. It holds that a deep estrangement of 'Chechnya' and 'Chechens' from 'Russia' was created through a collective and intersubjective (re) construction of this territory and this group of people. This made war acceptable in the first place, and produced a new but enduring blindness to the suffering of Chechnya and Chechens in Russia.

The first post-Soviet conflict over Chechnya, which erupted into full-scale war in 1994, was initially represented as a local separatist conflict. On the Russian side a primary reason for going to war was given as preventing the new Russian Federation from unravelling along the pattern of the Soviet Union.[7] 'Chechnya' was not detailed as a threat to 'Russia' in any substantive way before the war was launched.[8] In Chechnya, the leadership headed by General Dzhokhar Dudayev mobilized the population around primarily nationalist slogans as part of the build-up to the war, and the claim that Chechens could not survive under Russian rule acquired resonance among the Chechen population as the war was fought.[9] During the First Chechen War and the ensuing interwar years, Islam came to acquire a more prominent role in Chechen society, particularly among certain warlords who turned to Radical Islam. Their statements increasingly presented 'Russia' as an 'infidel' enemy and as an existential threat to the Muslims of the North Caucasus.[10]

On the Russian side, representations of 'Chechnya' changed as well. During the interwar years, official statements depicted President Aslan Maskhadov's Chechnya as a partner and friend. When the Second Chechen War was launched in October 1999, that move was presented as a response to the September 1999 terrorist attacks in Moscow, Volgodonsk and Buynaksk, which were blamed on Chechens. According to the Russian leadership, Chechnya had become 'a huge terrorist camp'.[11] The war itself was labelled a 'counter-terrorist campaign'.

While these radical shifts in the representations of the 'Other' on *both* sides in the Russo-Chechen conflict need to be investigated in order to understand the sum of gross violence and terror associated with the Second Chechen War, this book tells only half the story. I do not seek to attribute all blame on the Russian side or to deny that atrocities were committed by the Chechen side. Atrocities were committed on both sides during the Second Chechen War. There is no doubt that the escalation of the conflict to such violent heights was the result of a reciprocal process. However, this book has a narrower focus. The puzzle it tries to solve is how this war came to make so much sense on the Russian side.

Based in the tradition of Critical Security Studies, the origin of conflict is understood not as the outcome of timeless structures, but as grounded in reflexive practices. Rather than the competition of existing sovereign states or ethnic groups, the constitution of collective identity provides much of the impetus behind conflict.[12] Michael C. Williams has formulated this standpoint as follows:

> This is not to say that empirical elements are unimportant, but such conflicts cannot simply be reduced to the competing interests of pre-given political

objects. They are about the creation of these objects, and the way in which different identities are constitutive of them.[13]

Guided by a post-structuralist reading of securitization theory, the book assesses Russian re-phrasing of Chechnya by analysing the process of naming and describing the Chechen threat in official language (Chapters 4 and 5), evaluating to what extent representations among key groups in Russian society resonate with these official representations (Chapters 7, 8 and 9), and what kind of policies and practices of war these representations legitimized (Chapters 10, 11 and 12). It covers the years from 1996 to 2001, with an emphasis on autumn 1999. This timespan captures Russian official representations of and policies on Chechnya during the period between the two wars (1996–1999) and then Russian representations of the Chechen threat during summer and autumn 1999, as well as the material practices undertaken against Chechnya until 2001 in what most reasonably can be called the Second Chechen War. The First Chechen War (1994–1996), as well as parts of several hundred years of Russo-Chechen relations (Chapter 6), will be re-visited several times, but not in depth.

Although it is the war against Chechnya that is presented, the book provides general insight into how the mobilization and legitimation of war comes about in Putin's Russia and to what effect. It shows how a re-phrasing of another group or territory as an existential threat is essential to making the exercise of violence widely accepted in Russian politics and, at the same time, how such a re-phrasing creates cohesion in the fragmented Russian polity. On the one hand, the making of an acceptable war in Russia is a much more complex social process than most accounts make it look like. It is not decided and grounded in one authoritarian leader, but is a collective and intersubjective endeavour in which many societal actors take part. On the other hand, broad and collectively articulated representations of existential threat trigger a re-articulation of the threatened 'Self' and serve to re-constitute and unite the Russian political community internally. During the crises in Ukraine, we again witnessed the making of an acceptable Russian war. The object subjected to war is different. This time it was the 'fascist' Ukraine that was projected as a threat to Russia. But the broad social process that made Russian military action against Ukraine seem necessary and legitimate to the Russian public was recognizable. For those of us who have studied Russian war-making over years, the counter-terrorist campaign against Chechnya was a laboratory, a test case of how war becomes acceptable in Putin's Russia.

This holds also today. When Russian military force was employed in Syria in 2015 and 2016, it was made reasonable through a broader social process. The projection of terrorism as an existential threat to Russia was the focal point in this process. As put by Putin in his 2015 state of the nation speech, 'Russia has long been at the forefront of the fight against terrorism. This is a fight for freedom, truth and justice, for the lives of people and the future of the entire civilisation.'[14] While this looks like a radical shift away from Ukraine and the West as Russia's contemporary radical Others, it still has the effect of uniting, strengthening and re-constituting the Russian Self. Moreover, the articulation of international terrorism as the prime existential threat to Russia is powerful, credible and effective because it resonates with broad and enduring discursive patterns in Russian society. When Putin told the story of Russia's righteous fight against terrorism, the terrorist acts that sprang from the Chechen wars made up the foundation of his claims. A straight line was drawn from the Chechen threat to the present-day terrorist evil:

> We know what aggression of international terrorism is. Russia faced it back in the mid-1990s, when our country, our civilian population suffered from cruel attacks. We will never forget the hostage crises in Budennovsk, Beslan and Moscow, the merciless explosions in residential buildings, the Nevsky Express train derailment, the blasts in the Moscow metro and Domodedovo Airport.... It took us nearly a decade to finally break the backbone of those militants. We almost succeeded in expelling terrorists from Russia, but are still fighting the remaining terrorist underground. This evil is still out there. Two years ago, two attacks were committed in Volgograd. A civilian Russian plane was recently blown up over Sinai.... The militants in Syria pose a particularly high threat for Russia. Many of them are citizens of Russia and the CIS countries. They get money and weapons and build up their strength. If they get sufficiently strong to win there, they will return to their home countries to sow fear and hatred, to blow up, kill and torture people. We must fight and eliminate them there, away from home.[15]

Critical studies of Russia and terrorism

This study can be placed in the social constructionist camp. I believe that neither the threat nor the character of the Second Chechen War was determined by the nature of things.[16] It is not the aim of this book to argue that there was no Chechen threat, nor any threat from Radical Islamic fighters: determining the magnitude of the Chechen or Radical Islamic threat is scarcely feasible, and it is not my concern here.[17] The intention is rather to study how representations of 'Chechnya' in Russia have changed, how 'Chechnya' has been given a new meaning, and how this has influenced the means deemed legitimate for dealing with this Russian republic. Quite a few international *jihadi* fighters took part in the first post-Soviet war in Chechnya; the numbers participating in the Second Chechen War were not necessarily much higher. However, this fact was not spoken about during the First Chechen War, and the representation of 'Chechnya' prevalent in 1996 made negotiation and peace possible. In contrast, articulations of the Chechen enemy in Russia and of the Russian enemy among the Chechen insurgents during the Second Chechen War militated against such a solution.

The counterfactual reasoning which guides this book is that, if representations of Self and Other on each side of the Russo-Chechen conflict had been different, then different policies and practices would have been possible. In many ways, the whole book is an exploration of how discursive practices matter and work in making war and violence acceptable. But I do *not* suggest that acceptance of the Second Chechen War was an inevitable outcome of attempts by the Russian leadership to make it so. Rather, I point out how representations negating the version of Chechnya as a terrorist threat *could have* emerged in Russia to make the war unacceptable. This is an important point to make, given that it is difficult to imagine anything *but* a war-prone Russia in today's situation. Moreover, it renders the overall approach of this book both critical *and* constructive, in the sense that it will not only reveal how war becomes acceptable, but also indicate how war can be replaced by peaceful interaction.

With this approach, the current book is part of a broader endeavour to address and understand international and national security through discursive approaches.[18] Discursive approaches have proved particularly fruitful in analysing Western responses to terrorism, presumably the gravest threat in our time.[19] The current volume dovetails with a whole row of studies on another war that was midwifed through the discourse on terrorism at the beginning of this century, namely the war on Iraq.[20]

The main empirical contribution of this book is that it provides an in-depth, discursive analysis of Russia's version of the War on Terror. It enquires into how the security notion of terrorism is locally crafted in a non-Western setting and uncovers what the terrorist threat means to *Russian* politicians and publics. In this, the book answers Hagmann's[21] call to investigate the local, time and context specifics of security concepts, such as terrorism, and takes this agenda beyond the analysis of European and North American security issues, which dominates the field.

While too little light has been shed on Russia's version of the War on Terror in Western scholarship, Russian security politics are also seldom subjected to critical perspectives of a scholarly nature.[22] It is fair to say that this book, together with Bacon and Renz's study of securitization in Russia from 2006 and Aglaya Snetkov's recent book, introduces systematic and discursive approaches to understand the conflict in and over Chechnya.[23] There are several excellent books on the Chechen wars with rich empirical accounts.[24] But many of these are weak on theoretical concepts, and there is a general disregard of social theory. Some of these accounts even present the difference between 'Chechnya' and 'Russia' as an eternal and given fact, thereby further reifying a divide that in reality is constantly in the making.[25]

This endeavour is fundamentally critical. Adopting the post-structuralist approach of deconstruction was a conscious choice.[26] The post-structuralist approach aims to expose the processes of power and 'Othering' that are embedded in discursive structures and can be used to show how categories of our time become tools for oppression. Specifically, this book aims to show how the category of terrorism has produced a Chechen threat that Russian politicians so often give the impression of merely describing.

To make my personal standpoint clear, I believe that the imprint of the terrorist discourse on Russian–Chechen relations was a 'bad thing'. It has rendered concepts such as 'negotiation' and 'reconciliation' alien, and has legitimized the widespread use of violent emergency measures in Chechnya and the wider Northern Caucasus to this day. Not only has this created an even greater divide between Chechnya/Chechens and Russia: it will also pose an enormous challenge for social cohesion in the multi-ethnic and multi-confessional Russian state in years to come.

Moreover, the concern that the War on Terror has legitimized breaches of human rights and triggered a process of legal backsliding in several Western countries can be doubled in the case of Russia.[27] Although Russia after the fall of the Soviet Union adopted a liberal and democratic constitution (as well as a full set of new laws to detail this constitution) and signed all relevant human rights conventions, liberal laws and the protection of fundamental human rights do not have deep roots in Russia. This book studies how anti-terrorist discourse has shaped Russian approaches to Chechnya and Chechens, and in the final event suggests that the counter-terrorist campaign contributed to thwart the budding legal regime for the protection of basic human rights in Russia.

On the more general level, it proposes that the classification of 'terrorist' and the prominence of this classification in security language worldwide have increased the legitimacy of violence at the beginning of the twenty-first century. The prominence of this classification has not only opened the possibility for many leaders to launch violent responses against those classified as terrorists – it has also changed the dynamics of already ongoing conflicts, by allowing new and often extra-legal practices of war and excluding the possibility of peaceful solutions. Chechnya is merely one example.[28]

The process of securitization reconsidered

To uncover the social process that makes war acceptable, the book is informed by securitization theory but foregrounds the web of meaning and representation between a myriad of actors in society to unearth the contents – and changes – in how war is articulated and carried out with public consent. This matters not only for the question of how war becomes acceptable, but also for the very practices through which the war is fought: the emergency measures that are enabled in a discourse of existential threat. The social process that enables the legitimate undertaking of violent practices spring from an accumulation of statements that construct a sharp boundary between the Other as an existential threat and the threatened Self. A core theoretical argument advanced throughout the book is that the construction of such boundaries for acceptable violent action takes place through an *intersubjective* process. Not only official statements but also historical narratives, as well as those voiced by groups looked upon as 'audience' to official speech, contribute to making violent practices acceptable. In the case of the Second Chechen War, the communities of journalists, experts and parliamentarians contributed heavily to, and even initiated, the discourse of existential threat that made the war a legitimate undertaking.

This approach to understanding how war becomes acceptable draws on key concepts in securitization theory in a selective yet deepening way. It embeds securitization theory more firmly in post-structuralism – a move that builds on the theories of Ole Wæver, but also one that revises Copenhagen School securitization theory in fundamental ways. Three key differences between Copenhagen School securitization theory and the theoretical approach developed in this book will be highlighted here and then elaborated upon in Chapter 2. First, when security is accentuated as part of a constant and continuing social construction of reality as post-structuralist discourse theory encourages us to do,[29] securitization becomes a gradual, intersubjective process, not an instant, individual and intentional event as in speech act theory.[30] In a post-structuralist reading, securitization is not one utterance by one actor, but is produced over time in multiple texts that *represent* something as a threat. It is a result of an intersubjective struggle through texts emanating from 'securitizing actors' *and* 'audience' over what level of difference and danger to attach to something. The core of the process of securitization is a securitizing narrative that draws on and interacts with discursive structures and materializes in concrete emergency measures. This is a very different process from the actor-centric process suggested by Buzan, Wæver and de Wilde, which indicates a clear sequence, starting with a 'securitizing actor' that securitizes towards a 'significant audience' via a *speech act* in the Austinian 'once said, then done' (illocutionary) way.[31]

Second, in terms of the 'outcome' of an actor–audience agreement on something as an existential threat, Copenhagen School securitization theory merely states that this leads to the endorsement of emergency measures 'beyond rules that otherwise have to be obeyed'.[32] It gives no clue about what kinds of emergency measures are actually enabled in a process of securitization. In the post-structuralist perspective, meaning and materiality go hand in hand. Material practices are seen as intertwined with and complementing linguistic practices in the way proposed by Michel Foucault.[33] A post-structuralist framework for the study of securitization can therefore help us theorize the concrete material emergency measures that are enabled in a discourse of existential threat. When discourse is seen as a supra-concept which includes both significative practices and material practices, the statements that bring something into being as an existential terrorist threat are taken to acquire a logical and legitimate expression in material emergency measures. This book discusses not only how a new representation of the Chechen threat made it possible to launch the Second Chechen War, but also how the labelling of Chechnya (and Chechens) as a terrorist threat enabled the introduction of a whole series of counter-terrorist measures and practices against Chechnya (and Chechens) that might otherwise have been seen as illegitimate.

Finally, Copenhagen School securitization theory disregards the social effects that a process of securitization might have for the 'referent object' – the social entity that is said to be threatened. As McDonald suggested, 'a broader approach to the construction of security also entails a focus on how political communities themselves are constituted'.[34] Post-structuralist Self/Other literature encourages us to expand the focus of study beyond the (re)construction of threat (Chechnya) to include the (re)construction of the Self, the referent object (Russia). Based in a post-structuralist reading of securitization theory, this study will reveal where and how the hefty re-articulation of Russian identity associated with the Putin regime started.[35]

Second generation securitization theory post-Copenhagen School

This book does not only revise and develop Copenhagen School securitization theory, it also contributes to expand its applicability and engages the growing body of literature referred to as second generation securitization theory post-Copenhagen School.[36] In line with the Critical Security Studies agenda of introducing a greater range of issues, theoretical perspectives and methodological approaches, securitization theory was developed by Ole Wæver and Barry Buzan primarily to open up the study of security to a wider spectrum of issues beyond traditional military threats and to broaden the study of security, taking into consideration security by actors and referent objects other than the state.[37] While upholding the critical agenda, this study brings securitization theory back to the core of security studies by using it to understand how violence and war become acceptable in a state – Russia.[38] In this way, the book contributes to filling a gap in critical approaches to security identified by Hagmann: namely, that such approaches 'tend to focus ever more closely on political practices that have little bearing on *statist* foreign policymaking and thus interstate relations in the classic sense'.[39]

While Hagmann reworks securitization broadly into a framework to analyse how representations of danger are seized upon to justify distinct types of foreign policy strategies, this book will develop a framework to study the most dramatic of such strategies, namely war. Hagmann is uncomfortable with post-structuralism's focus on radical binary oppositions when theorizing the production of identity and consequently securitization theory's focus on posing threats as 'existential', because he wants to understand foreign policy production in general. But for the purposes of understanding how *war* becomes acceptable, the focus on radical Otherness and existential threat is crucial.[40] The possibility of launching violent measures on a scale such as war against an object, a territory or a social group hinges critically on representations of radical Otherness. Similarly, while Hagmann[41] draws on securitization theory *despite* the strong emphasis on the way in which naming threats as existential gives way to extraordinary – i.e. norm-breaking – powers, this book draws on it *because* of this emphasis on extraordinary force. War and violence is necessarily dramatically norm breaking, particularly when it is levelled by the state against its own citizens, as in the case studied in this book. How such extraordinary force is enabled by narratives of extreme difference and danger is the key to understanding how war becomes acceptable.

In many ways, the current endeavour joins efforts with Donnelly who has studied the war in Iraq through the prism of securitization.[42] As does Donnelly, I want to work in the broader framework of the Copenhagen School approach but shift 'attention away from the speech act as the dominant action towards a larger intersubjective activity that may involve multiple players'.[43] Yet our attempts at developing securitization as a framework to understand our specific cases of war produce different results. Donnelly stays on the level of state leadership and reworks securitization theory through Wittgensteinian insights to explore the different kinds of rules that exist when security is spoken, in particular the *constraints* of spoken words on actions. I explore and develop securitization theory as an analytical tool for understanding how war becomes acceptable via post-structuralist insights, with particular emphasis on how the 'audience' within a state contributes to such legitimation and how this legitimizes and *enables* violent practices.[44] This framework implies studying the full cycle of war as a securitization process, how war is prepared and made reasonable in language and the concrete implementation of this language in material practices such as bombing, cleansing, torture and murder. While securitization theory has inspired hundreds of studies, few of them investigate concrete emergency measures and provide detailed empirical examples of the material practices that securitizations enable. The failure to move beyond language is also a critique that is constantly levelled against post-structuralist and discourse theoretical contributions to security studies. Therefore, this book should be a welcome contribution to both of these bodies of literature.

By advancing a post-structuralist version of securitization theory, the book also contributes to the already long-running scholarly debate on this theory itself.[45] Considerable attention has been devoted to the claim that the Copenhagen School approach builds on two separate meta-theoretical convictions – neorealism and post-structuralism – something which gives rise to several contradictions and tensions. The debate has triggered efforts to specify and develop the theory into more distinct and coherent variants or views of securitization. Most notably, Balzacq[46] champions a 'sociological' version and Stritzel[47] suggests an 'externalist' version of securitization theory; simultaneously, they bracket what they call potential 'philosophical' or 'internalist' variants associated with post-structuralist traditions.

This book takes issue with Balzacq's attempt to dismiss the relevance of post-structuralism for theorizing the process of securitization, by mistaking it for Austinian speech act theory. According to Balzacq, 'those working in a post-structuralist tradition believe in a social magic power of language, a magic in which the conditions of possibility of threats are internal to the act of saying "security"'. This 'philosophical' variant, according to Balzacq '*ultimately* reduces security to a *conventional procedure*'.[48] Although Wæver indeed also seems to conflate the two in some of his writings, fundamental post-structuralist insights actually fly in the face of Austinian speech act theory. Most importantly, and as I will return to in the next chapter, the post-structuralist concept of discourse is an inherently social and intersubjective concept and stands in direct contradiction to the notion that 'the word ... is the act ... by saying it something is done'.[49]

In fact, the post-structuralist reading of securitization theory elaborated in this book seems to have more in common with the 'sociological' variant that Balzacq advances. Also a post-structuralist version sees securitization as a 'process that occurs within or as a part of a configuration of circumstances, including the context, the psycho-cultural disposition of the audience'.[50] A post-structuralist version considers the *discursive terrain* that any securitizing attempt feeds on and is launched into (see Chapters 2 and 6 of this book). However, it will emphasize the empowerment of the referent object at the expense of the thing that is said to threaten *as a consequence of a process of securitization* rather than emphasize 'the power that both the speaker and listener bring to the interaction', which the sociological variant is said to do.[51] A post-structuralist version will view securitization as a non-intentional process, because agency is placed in discursive practices and not in people. It therefore departs from the sociological version that sees securitization as both intentional and non-intentional. The conception of the audience, however, in many ways coincides. In the post-structuralist reading, the audience is *not* viewed as a 'formal-given-category, which is often poised in a receptive mode', as suggested by Balzacq, but rather more as an 'emergent category that must be adjudicated empirically'[52] just like in the sociological variant. As I will return to in the next chapter, the audience in the post-structuralist perspective is conceived of as a potential field into which the securitizing attempt is launched. The malleable yet fixed quality of discourses and the struggles between them means that the discursive reception of the securitizing attempt in the 'audience' is undecided. There is always room for change and appropriation of the securitizing narrative: it is not as if the 'audience' has already made up its mind before the transaction takes place.

Stritzel also offers a 'theorization of securitization as a specific conceptualization of discourse dynamics',[53] but grounds his re-reading of securitization theory in Norman Fairclough's critical discourse analysis.[54] With this move, Stritzel, as he notes, works in the tradition of problematizing the 'power in discourse', which is concerned 'with socio-political resources and power positions of actors, their political struggles and processes of authorization as specific moments in time to create, challenge, change or amend existing meaning structures, potentially establishing new discursive hegemonies as particular points in time'. By contrast, this book stays within the post-structuralist tradition of problematizing the 'power of discourse', where agency is located within discursive structures and through discursive changes rather than in social actors.[55] While Stritzel studies language *and* political constellations, this study focuses on language and how political constellations are created through discourse. In concrete terms, the focus is not on how the position as Prime Minister/President of Russia authorized Putin to speak security, but rather on how his speeches contributed to redrawing the boundaries between Russia and Chechnya in a fashion that empowered Russia at the expense of Chechnya and, in turn, authorized Putin as *the* speaker for Russia. The current explication of securitization theory does dovetail with Stritzel's in that I want to contextualize a securitization and follow how it travels from locale to locale. But again, while Stritzel broadens and follows the securitizations in world affairs, my project has been to follow how the discourse of danger travels within a state: from newspaper pages to presidents and parliamentarians and down to policemen and soldiers, all the time enabling and legitimizing violent emergency practices.

Outline of chapters

Having briefly introduced the essence of a post-structuralist reading of securitization and drawn up the boundaries in relation to the Copenhagen School approach as well as those of core second-generation securitization theorists, Chapter 2 next explicates and details securitization as a framework to analyse how war becomes acceptable. The framework has been developed to address the specific case of how the Russian counterterrorist campaign against Chechnya became acceptable. It is hoped that it also provides some more general conceptual tools for researching the processes through which leaders and their publics come to understand some violent conflicts as difficult to engage in and others as logical, even inescapable. Chapter 3 translates conceptual tools into working tools by presenting the research method employed, as well as the sources relied upon in this book.

The first empirical chapter of this book, Chapter 4, begins by re-visiting the interwar period (1996–1999). It aims to show that Chechnya has not always figured as Russia's radical Other, nor does it need to do so. Tracing official representations of 'Chechnya' and 'Russia' as well as the policies and practices pursued by the Russian leadership in relation to Chechnya in these years shows that a 'discourse of reconciliation' dominated. The Chechen issue was desecuritized in official Russian language, enabling negotiation and cooperation. Chapter 5 then moves on in time and investigates the official representations of 'Chechnya' and 'Russia' during spring, summer and autumn 1999. Official statements presenting Chechnya as an existential terrorist threat to Russia accumulated during this time; I present the details of this official securitizing narrative. The chapter concludes that the Second Chechen War was justified well in advance by the Russian leadership – not after the fact, as with the First Chechen War.

Chapter 6 revisits the Russian discursive terrain, with an analysis of the multitude of historical representations of 'Chechnya' and 'Russia' into which the official securitizing narrative was launched. While the need for a new war against Chechnya was argued for at length by the Russian leadership, such a discursive terrain offers both possibilities and constraints. And indeed, the sharp demarcation between 'Russia' and the 'Chechens'/'Chechnya' is revealed as having been centuries in the making, resonating strongly with the new official securitizing narrative.

Chapter 7 then casts the net even wider by investigating representations of 'Russia' and 'Chechnya' in statements of the Russian political elite holding or campaigning for seats in the Federal Assembly during autumn 1999. Here the premise is that representations by groups presented merely as an 'audience' in

Copenhagen School securitization theory *could* have discarded the 1999 official securitizing narrative, even if it was well argued and resonated well with the Russian discursive terrain. In Chapters 8 and 9, expert and media texts are examined. They too detail and even expand the representation of 'Chechnya' as different and dangerous and that of 'Russia' as a righteous defender. The core argument throughout Chapters 7, 8 and 9 is that the process that brought Chechnya into being as an existential terrorist threat was not the achievement of Prime Minister Putin in isolation: it was a collective and intersubjective endeavour. The words of the political elite beyond the Kremlin, of the experts and the journalists, played a key role in transmitting the new core understandings of 'Chechnya' and 'Russia' to the broader Russian public. When the Russian ground offensive into Chechnya started in October 1999, the Second Chechen War had become an acceptable undertaking.

The three final empirical chapters, Chapters 10, 11 and 12, move from linguistic representations and on to investigating the material practices of war. Chapter 10 shows how the urgent security situation entailed in the securitizing narrative immediately became translated into the endorsement of emergency measures proposed by the Russian leadership. In line with the post-structuralist approach of this book, practical enactments of representations are given more attention than such initial formal endorsement. The practices that served to 'seal off' Chechnya and Chechens from Russia are presented. These practices were both logical and legitimate, given the new one-sided classification of Chechnya; and their enactment contributed to reify this classification with yet another layer. Chapter 11 examines the intensive and repeated bombing of Chechen territory, which was on a par with that of the First Chechen War, from September 1999 onward. Finally, the violent practices undertaken against the population of Chechnya in connection with the efforts to 'cleanse' this Russian republic of terrorists during the ground offensive from October 1999 onward are discussed in Chapter 12. A core concern throughout these Chapters 10, 11 and 12 is to show how language functioned to legitimize violent practices from the outset until they were carried out and how such material practices in turn served to constitute, confirm and cement the identity of Chechnya and Chechens as a terrorist threat to Russia.

The concluding Chapter 13 begins by presenting some general claims about securitization and war. It then summarizes the core findings of the empirical case studied throughout the book in order to draw out some broader perspectives on the functioning and consequences of the counter-terrorist campaign for Russia as a state. As a codicil, I present the life of Chechen President Aslan Maskhadov as an allegory of the Second Chechen War.

Notes

1 C.W. Blandy. 2000. *Chechnya: Two federal interventions. An interim comparison and assessment.* Camberley: Conflict Studies Research Centre, Royal Military Academy, Sandhurst, 46.

2 In January 1995 only 22.8 per cent of the Russian population was for the use of armed force to solve the conflict in Chechnya, and 54.8 per cent was explicitly opposed. This mood was confirmed in January 1997 by strong support (67 per cent) for the Khasavy-urt Accord. By November 1999, 52 per cent were in favour of establishing constitutional order in Chechnya by use of the army (B.K. Levashov. 2001. *Rossiyskoye obshchestvo i radikal'nye reformy.* Moscow: Akademia, Russian Academy of Science, Institute of Social-Political Research, 850–852). Emil Pain has documented in figures the radical shift in terms of public acceptance for war against Chechnya (2005. The Chechen War in the context of contemporary Russian politics. In: R. Sakwa (ed.) *Chechnya from Past to Future.* London: Anthem Press, 67–78).

3 T.P. Gerber and S.E. Mendelson. 2002. Russian public opinion on human rights and the war in Chechnya. *Post-Soviet Affairs,* 18(4): 271–305.

4 For key texts presenting the classical Copenhagen School version of securitization, see O. Wæver. 1995a. Securitization and desecuritization. In: R.D. Lipschutz (ed.) *On Security.* New York: Columbia University Press, 46–86; O. Wæver. 1995b. Identity, integration and security: Solving the sovereignty puzzle in E.U. studies. *Journal of International Affairs,* 48(2): 289–431; B. Buzan. 1997. Rethinking security after the Cold War. *Cooperation and Conflict,* 32(1): 5–28; B. Buzan, O. Wæver and J. de Wilde. 1998. *Security: A new framework for analysis.* Boulder, CO: Lynne Rienner.

5 For key post-structuralist works on Self/Other see R.B.J. Walker. 1990. Sovereignty, Identity, Community: Reflections on the Horizons of Contemporary Political Practice. In: R.B.J Walker and S.H. Mendlovitz (eds.) *Contending Sovereignties: Redefining political community.* Boulder: Lynne Rienner, 159–185; S. Dalby. 1988. Geopolitical discourse: The Soviet Union as Other. *Alternatives,* 13(4): 141–155; M. Dillon. 1990. The alliance of security and subjectivity, *Current Research on Peace and Violence,* 13(3): 101–124; W.E. Connolly, 1991. *Identity/Difference: Democratic negotiations of political paradox.* Ithaca, NY: Cornell University Press; D. Campbell. 1992. *Writing Security.* Minneapolis, MN: University of Minnesota Press; I.B. Neumann. 1996. *Russia and the Idea of Europe.* London: Routledge; R.L. Doty. 1996. *Imperial Encounters.* Minneapolis: University of Minnesota Press; L. Hansen. 2006. *Security as Practice.* New York: Routledge.

6 Former Prime Minister Sergey Stepashin revealed that Russia made its plans to invade Chechnya six months *before* the events that are thought to have triggered the Second Chechen War: the summer 1999 incursion into Dagestan and the apartment bombings in Russian cities. ('Russia planned Chechen war before bombings', *The Independent,* 29 January 2000.)

7 C. Gall and T.D. Waal. 1997. *Chechnya: A small victorious war.* London: Pan Original; A. Lieven. 1998. *Chechnya: Tombstone of Russian power.* New Haven, CT: Yale University Press; J.B. Dunlop. 1998. *Russia Confronts Chechnya: Roots of a separatist conflict.* Cambridge: Cambridge University Press.

8 C. Wagnsson. 2000. *Russian Political Language and Public Opinion on the West, NATO and Chechnya.* Stockholm: Akademitryck AB Edsbruk.

9 J. Wilhelmsen. 1999. *Conflict in the Russian Federation: Two case studies, one Hobbesian explanation.* NUPI Report 249. Oslo: NUPI.

10 D. Trenin and A. Malashenko. 2004. *Russia's Restless Frontier: The Chechnya factor in post-Soviet Russia.* Washington, DC: Carnegie Endowment for International Peace; J. Wilhelmsen. 2005. Between a rock and a hard place: The Islamisation of the Chechen separatist movement. *Europe–Asia Studies,* 57(1): 35–59; C. Moore and P. Tumelty. 2009. Unholy alliances in Chechnya: From Communism and nationalism to Islamism and Salafism. *Journal of Communist Studies and Transition Politics,* 25(1): 73–94. R. Dannreuther. 2010. Islamic radicalization in Russia: An assessment. *International Affairs,* 86 (1): 109–126.

11 Prime Minister Vladimir Putin, cited in 'Tret'ya otechestvennaya?', *Monitor,* 15 September 1999.

12 K. Krause and M.C. Williams. 1997. From strategy to security: Foundations of critical security studies. In: K. Krause and M.C. Williams (eds.) *Critical Security Studies*. London: Routledge, 33–61.
13 Ibid., 47.
14 Presidential Address to the Federal Assembly December 3, 2015, available at Kremlin.ru (http://en.kremlin.ru/events/president/news/50864).
15 Ibid.
16 I locate this study in the social construction tradition on the basis of Ian Hacking's proposition that what unites various types of social construction work is a stand against inevitability (I. Hacking. 1999. *The Social Construction of What?* Cambridge, MA: Harvard University Press).
17 Threats cannot be understood as objectively given and cannot be studied as such. They are determined not by the nature of things, but through discourse. This is not to say that there is no substance to the threat (indeed – heinous, violent acts aimed at civilians may be committed). It is the *concept* (of terrorism) as a threat that is viewed as socially constructed.
18 For overviews, see K.M. Fierke. 2007. *Critical Approaches to International Security*. Cambridge: Polity; B. Buzan and L. Hansen. 2009. *The Evolution of International Security Studies*. Cambridge: Cambridge University Press; C. Peoples and N. Vaughan-Williams. 2010. *Critical Security Studies: An introduction*. London: Routledge.
19 Notable examples include J. Der Derian. 2005. Imaging terror: Logos, Pathos and Ethos. *Third World Quarterly*, 26(1): 23–37, on the elusiveness of the concept of terrorism; C. Weber. 2006. An aesthetics of fear: The 7/7 London bombings. *Millennium*, 34(3): 683–710, on how the aesthetics of fear were politically mobilized in the case of the London 7/7 bombings; O. Wæver. 2006. What's religion got to do with it? Terrorism, war on terror, and global security. Keynote lecture at the *Nordic Conference on the Sociology of Religion*. Aarhus, 11 August 2006, on the securitization processes at work in the rhetorical battles between George Bush and Osama bin Laden. See also B. Buzan and O. Wæver. 2009. Macrosecuritization and security constellations: Reconsidering scale in securitization theory. *Review of International Studies*, 35(2): 253–276. Several titles in the Routledge Critical Terrorism Studies are written from a related perspective: see R. Jackson, M. Breen Smyth and J. Gunning (eds.) 2009. *Critical Terrorism Studies: A new research agenda*. London: Routledge; B. Brecher, M. Devenney and A. Winter (eds.) 2010. *Discourses and Practices of Terrorism: Interrogating terror*. New York. Routledge; A.C. Stephens and N. Vaughan-Williams (eds.) 2009. *Terrorism and the Politics of Response*. New York: Routledge; R. Jackson, E. Murphy and S. Poynting (eds.) 2011. *Contemporary State Terrorism: Theory and practice*. London: Routledge; M. Thorup. 2010. *An Intellectual History of Terror: War, violence and the state*. London: Routledge; J. Holland. 2013. *Selling the War on Terror: Foreign policy discourses after 9/11*. New York: Routledge.
20 R. Jackson. 2005. *Writing the War on Terrorism: Language, politics and counterterrorism*. Manchester: Manchester University Press; S. Croft. 2006. *Culture Crisis and America's War on Terror*. Cambridge: Cambridge University Press; R.R. Krebs and J.K. Lobasz. 2007. Fixing the meaning of 9/11: Hegemony, coercion, and the road to war in Iraq. *Security Studies*, 16(3): 409–451; J. Holland. 2012. *Selling the War on Terror: Foreign policy discourses after 9/11*. New York: Routledge; F. Donnelly 2013. *Securitization and the Iraq War: The rules of engagement*. New York: Routledge.
21 J. Hagmann. 2015. *(In-)Security and the Production of International Relations: The politics of securitization in Europe*. London: Routledge: 3, emphasis added.
22 M. Galeotti. 2010. *The Politics of Security in Modern Russia*. Farnham: Ashgate.
23 A. Snetkov. 2015. *Russia's Security Policy under Putin: A critical perspective*. London: Routledge. The present book in many ways complements Snetkov's study

(2015). Her focus is on Russia's security policy under Putin from 2000 to 2014 as analysed through the prism of official Russian discourse on Chechnya. The current volume also scrutinizes the Russian discourse on Chechnya, but over a shorter time-span and in more depth. While Snetkov shows how official discourses are intertwined in the domestic–foreign policy nexus, this study focuses on the ways in which domestic constituencies beyond or beneath the official level of politics feed into discourses of existential threat in Russia. In this it also departs from Bacon and Renz's approach, which sees securitization as a narrow and intentional act undertaken by the political leadership across a range of policy areas to create an autocratic system (E. Bacon and B. Renz, with J. Cooper. 2006. *Securitizing Russia: The domestic politics of Putin*. Manchester: Manchester University Press). By contrast, this book focuses inward on the one event of how the Second Chechen War became an acceptable undertaking and moves beyond the level of general security policy to investigate and present the detail of brutal warring practices during the Second Chechen War.

24 Gall and Waal 1997; V. Tishkov. 1997. *Ethnicity, Nationalism and Conflict in and after the Soviet Union: The mind aflame*. London: SAGE; Dunlop. 1998; Lieven 1998; S. Smith. 1998. *Allah's Mountains. The battle for Chechnya*. London: Tauris Parke; A. Nivat. 2001. *Chienne de Guerre. A woman reporter behind the lines of the war in Chechnya*. New York: Public Affairs; R. Seely. 2001. *Russo-Chechen Conflict, 1800–2000*. London: Frank Cass; M. Evangelista. 2002. *The Chechen Wars. Will Russia go the way of the Soviet Union?* Washington, DC: Brookings Institution Press; V. Tishkov. 2004. *Chechnya: Life in a war-torn society*. Berkeley, CA: University of California Press; Trenin and Malashenko 2004; A. Meier. 2005. *Chechnya. To the heart of a conflict*. New York: W.W. Norton; R. Sakwa. 2005. *Chechnya from Past to Future*. London: Anthem Press; M. Gammer. 2006. *The Lone Wolf and the Bear. Three centuries of Chechen defiance of Russian rule*. London: Hurst; M.D. Toft. 2006. Issue indivisibility and time horizons as rationalist explanations of war. *Security Studies*, 15(1): 34–69; J. Hughes. 2007. *Chechnya: From nationalism to Jihad*. Philadelphia, PA: University of Pennsylvania Press; J. Russell. 2007. *Chechnya: Russia's 'War on Terror'*. London: Routledge; E. Gilligan. 2010. *Terror in Chechnya. Russia and the tragedy of civilians in war*. Princeton, NJ: Princeton University Press.

25 For an excellent treatment and critique of the 'historicist' approach to the Chechen conflict, see Hughes (2007).

26 M. Foucault. 1972. *The Archaeology of Knowledge and the Discourse on Language*. New York: Pantheon Books; J. Derrida. 1967. *Of Grammatology*. Baltimore, MD: The Johns Hopkins University Press; J. Kristeva. 1980. *Desire in Language: A semiotic approach to literature and art*. New York: Columbia University Press.

27 As a response to this concern, the UN established a Special Rapporteur on Terrorism and Human Rights in 2007.

28 For other examples, see A. Houen (ed.) 2014. *States of War since 9/11: Terrorism, sovereignty and the War on Terror*. London: Routledge.

29 E. Laclau and C. Mouffe. 1985. *Hegemony and Socialist Strategy: Towards a radical democratic politics*. London: Verso; J.L. Austin. 1962. *How to Do Things with Words*. Oxford: Clarendon Press.

30 The notion of *speech act* is taken from Austin (1962), who argued that statements can be used to perform an action, as with the statement 'I do' in a marriage ceremony. Austin called these 'performative speech acts'. The Copenhagen School paraphrases the illocutionary act for its definition of securitization (Wæver 1995a: 55).

31 Buzan, *et al.* 1998.

32 Ibid., 25

33 M. Foucault. 1995. *Discipline and Punish: The birth of the prison*. New York: Vintage Books.

34 M. McDonald. 2008. Securitization and the construction of security. *European Journal of International Relations*, 14(4): 563–587, 565.

35 For an account that sheds light on this process of articulating Russia as a strong state over the long timespan 2000 to 2014, see Snetkov 2015.
36 Key contributions to this literature are H. Stritzel. 2007. Towards a theory of securitization: Copenhagen and beyond. *European Journal of International Relations*, 13(3): 357–383; H. Stritzel. 2014. *Security in Translation: Securitization theory and the localization of threat*. London: Palgrave; R. Taureck. 2006. Securitization theory and securitization studies. *Journal of International Relations and Development*, 9(1): 52–61; T. Balzacq. 2005. The three faces of securitization: Political agency, audience, and context. *European Journal of International Relations*, 11(2): 171–201; T. Balzacq (ed.). 2011. *Securitization Theory: How security problems emerge and dissolve*. Abingdon: Routledge; McDonald 2008; M.B. Salter. 2008. Securitization and desecuritization: Dramaturgical analysis and the Canadian Aviation Transport Security Authority. *Journal of International Relations and Development*, 11(4): 321–349; J. Vuori. 2008. Illocutionary logic and strands of securitization: Applying the theory of securitization to the study of non-democratic political orders. *European Journal of International Relations*, 14(1): 65–99; R. Floyd. 2010. *Security and the Environment: Securitization theory and U.S. environmental security policy*. Cambridge: Cambridge University Press; S. Croft. 2012. *Securitizing Islam: Identity and the search for security*. Cambridge: Cambridge University Press; the 2011 special issue of *Security Dialogue* 42 (4–5) on The Politics of Securitization; Donnelly 2013; and Hagmann 2015.
37 For overviews of CSS, see Krause and Williams 1997; Peoples and Vaughan-Williams 2010; C.S. Browning and M. McDonald. 2013. The future of critical security studies: Ethics and the politics of security. *European Journal of International Relations*, 19(2): 235–255.
38 While this means limiting the focus of the book to the mode of securitization which requires a radical separation of Self and Other i.e. national/societal security, it does not deny that several modes of securitization characterized by different logics (e.g. human security) may operate simultaneously, as suggested by R. Doty. 1998. Immigration and the politics of security. *Security Studies*, 8(2–3): 71–93.
39 Hagmann 2015: 3 emphasis added.
40 Ibid., 17–18.
41 Ibid., 18.
42 Donnelly 2013.
43 Ibid., 3.
44 For a useful distinction between critical constructivism (which Donnelly herself works from) and post-structuralism, see Donnelly (2013: 17–21).
45 Hansen's (2011) post-structuralist reading of securitization theory offers many suggestions similar to those presented here, but couched in a different language: she draws directly on Foucault, whereas I draw on a collective body of insight from various IR (international relations) scholars who employ a post-structuralist approach. See L. Hansen. 2011. The politics of securitization and the Mohammad cartoon crises: A post-structuralist perspective. *Security Dialogue*, 42(4/5): 357–369.
46 Balzacq 2005, 2011.
47 Stritzel 2007, 2014.
48 Balzacq 2011: 1.
49 Wæver 1995a: 55.
50 Balzacq 2011: 1.
51 Ibid., 1–2.
52 Ibid., 2.
53 Stritzel 2014: 39.
54 A confusing aspect of Stritzel's work is that he claims to build on Norman Fairclough's critical discourse analysis, but constantly invokes Laclau and Mouffe to explain and explicate his discursive securitization theory. Laclau and Mouffe largely

follow Foucault and arguably advance an *alternative* approach to discourse from Fairclough's.

55 Stritzel 2014: 39. The distinction between 'power in' and 'power of', which is referred to by Stritzel, is coined by A. Holtzscheiter. 2010. *Children's Rights in International Politics: The transformative power of discourse*. London: Palgrave Macmillan.

2   A theory on acceptable war

The theoretical framework that guides and structures this study is informed by securitization theory, Patrick Jackson's concept of legitimation (2006) and post-structuralist propositions.[1] Although these are different contributions to the IR literature and operate on different analytical levels, they complement each other and work in combination because they are all committed to the same 'philosophical wager', namely *mind–world monism*, the view that social objects do not exist independent of our ideas or beliefs about them.[2] In post-structuralist parlance, which has particularly influenced this book, the objects of our knowledge are neither objectively given, nor independent of our interpretations or language, but are products of our ways of categorizing the world. That is not to say that discourse has priority over non-discourse, that objects do not exist without thought or language – but 'that they could not constitute themselves as objects outside of any discursive condition of emergence'.[3] In line with this monist position, the theory of acceptable war is informed by a particular understanding of threat. The object subjected to war does not necessarily constitute a 'real' existential threat, but it is represented as such.

Copenhagen School securitization theory was chosen as a point of departure because the dynamics and concepts outlined in this theory seemed to capture the logic of what was happening in Russia when the Second Chechen War was launched. Indeed, according to Wæver, the Copenhagen School endeavour was precisely about uncovering 'what practitioners actually do in talking security'.[4] Copenhagen School securitization theory provides a conceptualization of a process that combines the onset of urgent securitizing talk which positions a particular issue as a threat to survival, acceptance of this representation among the audience, and the enabling of emergency measures 'beyond rules that would otherwise bind'.[5] When Vladimir Putin became Prime Minister in summer 1999, he clearly framed Chechnya as an existential threat to Russia in his speeches. His claim was so widely accepted in the Russian public that a war, the crudest of emergency measures, could be undertaken without protest. Such a war had been beyond the rules of the game, defined in terms of the domestic political debate, only half a year before.

Despite this intuitive, initial fit with the case at hand, Copenhagen School securitization theory is inadequate to catch the dynamic social process that goes on when war becomes acceptable. The making of an acceptable war is a much broader social process and has much wider societal ramifications than Copenhagen School securitization theory allows us to investigate. In the case of the Second Chechen War, the discursive process which established Chechnya as an existential threat was full of interaction across political leadership and public constituencies. It included many more voices than that of the Russian leader. In the course of this process Chechnya was given a new identity, but also Russian identity was re-constituted. The representations of Chechnya as an existential threat to Russia were a drastic break with the official representations only half a year before, yet this enmity had been centuries in the making. Finally, the radical re-articulations of Chechnya and Russia during 1999 were not confined to words: they manifested themselves quite literally in detentions, bombs, torture and killing. All this suggests the need to create a framework that pays more attention to inter-subjectivity, context, identity-change and materiality than Copenhagen School securitization theory does. Such a framework is within reach if we take the post-structuralist underpinnings of securitization theory more seriously.

This chapter will proceed in four sections. The next section outlines how post-structuralist ideas on policy production can be used to re-phrase securitization as a discursive process of legitimation. It includes an argument in favour of substituting a post-structuralist concept of discourse for the Austinian speech act as a core concept in the theory. Picking up on and specifying this broader outline of securitization as a discursive process of legitimation, I then in the second, third and fourth sections move on to show what post-structuralist insight *does* to securitization theory when elaborated to address the question of how war becomes acceptable. In these sections, I expound key concepts and relations in securitization under headings suggested by the Copenhagen School, namely *representations of existential threat, emergency measures and audience acceptance*,[6] offering a post-structuralist re-interpretation.

This will imply disrupting the sequence from representations by securitizing actors via audience acceptance to emergency measures as suggested by the Copenhagen School in order to benefit from post-structuralist insights on how significative and material discursive practices are intertwined and co-constitutive. By emphasising the productivity of discourse, the framework will go beyond re-interpretation to provide insights 'unreleased' in Copenhagen School securitization theory, while stripping it of extra-discursive concepts. Where possible the writings of Wæver, Buzan and de Wilde are revisited to find support for the conceptualizations offered.[7] In other places, second generation securitization literature is engaged to contrast or to find support. Research questions that will guide the empirical enquiry in this book are formulated, and several caveats issued as to the Russian case of securitization for war in 1999.

Securitization as a discursive process of legitimation

The core insight of Copenhagen School securitization theory is that issues, military as well as non-military, can become 'securitized' when 'securitizing actors' (for example, political leaders or pressure groups), by means of rhetorical strategies, elevate them to the status of an existential threat to a referent object (for example, individuals, the state or the environment) *and* when a significant audience accepts this representation of the issue.[8] This process generates endorsement for emergency measures (such as the use of military force, secrecy, additional executive powers) beyond otherwise binding rules. According to Buzan, Wæver and de Wilde 'the invocation of security has been the key to legitimizing the use of force, but more generally it has opened the way for the state to mobilize, or to take special powers, to handle existential threats'.[9] However, threats can also be de-securitized. Issues become de-securitized when they are shifted out of emergency mode and into the normal bargaining processes of the political sphere.[10]

In sum, there are three components in a process of securitization as suggested by the Copenhagen School. First, there is the identification by securitizing actors of something as an existential threat, and then there are two other components – effects on inter-unit relations (audience acceptance) and breaking free of rules; and emergency action.[11]

The general thrust of the argument underlying this description of the securitization process is in many respects in line with post-structuralist ideas of how policies are co-constituted by identities and rely on accounts that make sense of them and legitimize them as they are launched.[12] The post-structuralist stand is that policies are not a given response to an external reality to which the state (or other social actors) relates objectively, but are co-constituted by ideas or identities. As stated by Lene Hansen:

> foreign policies need an account, or a story, of the problems and issues they are trying to address: there can be no intervention without a description of the locale in which the intervention takes place, or of the people involved in the conflict.[13]

References to identities are necessary to represent and legitimize policies, but at the same time these identities are constituted and reproduced through the formulation of policies. This is why the term 'co-constituted' is used.

When the production of politics is understood in this way, the task for the analyst is to 'embrace a logic of interpretation that acknowledges the improbability of cataloguing, calculating and specifying "real causes" and concerns itself with considering the manifest political consequences of adopting one mode of representation over another'.[14] Studying politics, then, involves studying how some representations of reality become dominant discourses, and how problems, subjects and objects are constituted in these discourses that simultaneously indicate relevant policies to pursue.

The claim is not that such dominant representations *cause* certain policies or actions, but that they both open up and constrain the range of policies and actions that seem possible and *legitimate* to undertake.[15] With the help of Patrick Jackson, we can conceptualize this link between identity and policies more explicitly as one of legitimation. Jackson sees legitimation as:

the process of drawing and (re)establishing boundaries, ruling some courses of action acceptable and others unacceptable. Out of the general morass of public political debate, legitimation contingently stabilizes the boundaries of acceptable action, making it possible for certain policies to be enacted.[16]

'The process of drawing and (re)establishing boundaries' is here taken to be the continuous references to Self and Other that policy formulation implies and which, in turn, legitimates policy implementation.

Based on this understanding of how policies are produced and legitimized, a post-structuralist approach would imply treating securitization as a process through which a representation of something as an existential threat becomes dominant at the expense of other representations, and uncovering, in the course of research, the changing boundary between this identity and that given to the 'referent object' (I elaborate on this in the section 'Representations of existential threat'). These changing representations would not determine emergency action, but would condition the range of emergency measures political actors could undertake legitimately (I elaborate on this in the section 'Emergency measures'). In turn, the undertaking of such emergency measures against the something that is said to be threatening and in defence of the referent object would confirm and reinforce the new identity boundaries that were drawn up and legitimize the undertaking of emergency measures in the first place. From a post-structuralist perspective, the process of securitization is a fundamentally co-constitutive process and one that implies legitimizing concrete security policies and material practices. This reading suggests that securitization cannot be treated as a sequential, linear process from a securitizing move by an actor through audience acceptance to emergency measures, as suggested in Copenhagen School securitization theory. Rather it is a much more dynamic development; threat objects, referent objects, securitizing actors, as well as exceptional measures, are co-constituted within the securitization process. Moreover, alternative articulations of the Self/Other boundary which attach a lower level of threat to the Other can emerge and render emergency measures illegitimate. This would be a process of de-securitization.

Discourse, not Austinian speech act

According to Copenhagen School securitization theory, threats and security are determined through the *speech act*.[17] Understood as a speech act, 'security' means that the very identification, the articulation of words, that describe something as a security threat, is an act. To quote Wæver:

> security is not of interest as a sign that refers to something more real: the utterance itself (in original) is the act. By saying it, something is being done (as in betting, giving a promise, naming a ship). By uttering 'security', a state-representative moves a particular development into a specific area, and thereby claims a special right to use whatever means are necessary to block it.[18]

The weight given to words in this explication seems to match post-structuralists' foregrounding of language, but it is still not useful to theorize securitization as a *speech act* in the way that Wæver seems to do (as a self-referential practice, an *illocutionary* act in John Austin's vocabulary).[19]

First, it is an impossible attempt at reconciling relationalism with actor-centric understandings of social change. The insistence on working from a 'securitizing actor' that launches a 'securitizing attempt' towards a 'significant audience' via a *speech act* in the Austinian 'once said, then done' way contradicts the emphasis on intersubjectivity elsewhere in the theory.[20] The initial focus on actors implicit in Copenhagen School securitization theory is amplified with the adoption of the speech act in the illocutionary form. Not only is the securitizing actor projected as the driving force in the process, but his words are also accorded status as final and decisive. Thus, the audience is certainly not significant and the intersubjective process is lost.[21]

By adhering firmly to relationalism and placing agency not in an actor, but in the discursive practices that comprise a securitization, a post-structuralist approach offers a less contradictory theoretical framework and one that gives priority to intersubjectivity.[22] In comparison to Austinian speech acts in the illocutionary form, *discourses* are much more comprehensive – and they are inherently intersubjective.

Discourses are seen as structures of signification which construct social realities. The understanding of significative construction which informs most post-structuralist work is taken from the structuralist linguistics of Ferdinand de Saussure (1974). He held that language is not determined by the reality to which it refers – it should be understood as a system of signs, with the meaning of each sign determined by its relation to other signs. A sign is thus part of a structure together with other signs that it differs from, and it gains its specific value precisely from being different from other signs.[23] The assumption, prevalent in most post-structuralist IR work, that discourses are structured largely in terms of binary oppositions draws on the work of Jacques Derrida. According to Derrida, language is a system of differential signs and meaning is established not by the essence of a thing itself but through a series of juxtapositions, where one element is valued over its opposite. Binary oppositions are not neutral: they establish a relation of power such that one element in the binary is privileged.[24]

Discourses are made in a process of social interaction and are always textually interconnected; as such, they are a set of collectively articulated codes and intersubjectively embedded at the outset. Moreover, they are continuously conditioned by intersubjectivity – because, despite being highly structured, they are seen not as stable grids, but as open-ended, changeable and historically contingent.[25] This aspect of discourse implies that there is a *play of practice*, or *struggle* over which discourses should prevail. Whatever the label affixed by theorists of discourse, the main idea is that meaning can never be ultimately fixed – because, in ongoing language use, signs are positioned in various relations to one another so that they may acquire new meanings. This in turn entails constant struggles and negotiations in social contexts to fix and challenge the meaning of signs by placing them in particular relations to other signs. Some fixations of meaning become so conventionalized that we think of them as natural. Other fixations are always possible, but may become temporarily excluded by these hegemonic discourses.[26] Replacing the Austinian 'speech act' with post-structuralist 'discourse' enables us to move from a self-contained and definite core concept in the theory to an inherently intersubjective and process-oriented concept. As we shall see below, this move will elicit substantial re-conceptualizations of 'audience' and 'context' (I elaborate on this in the section 'Audience acceptance and breaking free of rules').

A second argument for replacing the Austinian 'speech act' with post-structuralist 'discourse' is that we get a better fit between theory and real world observations. If we consider what securitization would look like in the empirical world, a reasonable understanding would be that a 'securitizing attempt' consists of *a series* of utterances. Nothing can be constituted as an existential threat on a political arena through a 'speech act' in the 'once said, then done' sense. As Butler points out, the power of speech acts (not to be confused with Austinian illocutionary speech acts) lie in their *iterability*, i.e. they can be cited, recited and changed through such citation. It is only through iterability that utterances have transformative potential.[27] This makes it more appropriate to understand 'securitizing moves' as the onset or strengthening of a *discourse* that constructs something as an existential threat.

Ronald Krebs and Patrick Jackson have argued along these lines, saying that 'rhetorical innovation, while possible and even inevitable in the long run, is far less likely in the short run'.[28] One reason is that, even if discourses are never fully fixed, 'coherent political action would be impossible if rhetorical universes were in a state of continuous deep flux. Relative rhetorical stabilities must emerge to permit the construction of political strategies.' Further, making and distributing new representations takes time and effort. According to Krebs and Jackson, 'Arguments can prove powerful only when the commonplaces on which they draw are *already* present in the rhetorical field.'[29] Securitizing attempts, if they are to have any security effects, are thus not born in one rhetorical instance, but in a series of expressions that are innovative, yet bounded.

Finally, as regards the application of securitization to understand how war becomes acceptable, a discursive approach simply seems to have greater explicatory clout. The Second Chechen War and the acceptance of this violent undertaking by the broader Russian public cannot be thought of as a single authoritative act: it is better grasped as an evolving intersubjective process. The purpose of an enquiry informed by post-structuralist securitization theory is to study how this intersubjective process of securitization unfolds.

The next section details what must be considered the focal point within such a process of securitization, namely representations of existential threat. The account fleshes out several post-structuralist re-conceptualizations under this label and proposes how to translate these conceptualizations into concrete analytical tools to study an empirical case. I elaborate on how we should understand a 'securitizing move'[30] and introduce 'the securitizing narrative' as a template through which to study the linguistic aspects of securitization. The securitizing narrative caters for post-structuralist insights, in that it moves beyond the focus on representations of existential threat to include representations of the referent object, the changing boundaries between these two mutually constitutive identity constructions as well as the policy suggestions implicit in any securitizing argument. Building on the choice of substituting the Austinian speech act with a concept of discourse that is sensitive to the dynamics and changes in social relations, I also suggest 'scaling threat' so that it is possible to determine when a threat construction has passed the threshold of becoming an *existential* threat. Finally, I discuss how broadening the study of securitization to include the changing representations of referent object identity can yield insights into the effects that securitization may have for the (re-)constitution and consolidation of the society that is said to be threatened.

Representations of existential threat

Although securitization theory can be read as putting the 'securitizing actors' first, Buzan, Wæver and de Wilde actually state that 'one cannot make the actors of securitization the fixed point of analysis – the practice of securitization is the centre of analysis'.[31] This practice is the significative practice of giving something the identity of an existential threat to a referent object. Putting the practice of securitization at the centre of analysis means that investigating *representations* is the starting point of an empirical enquiry.

In practical terms, this focus on representations instead of actor entails searching texts (including images) for an *accumulation* of statements that identify something as an existential threat.[32] With the substitution of discourse for Austinian speech act, a 'securitizing move' or 'attempt' in the terminology of Buzan, Wæver and de Wilde is not one such statement, but many. Moreover, the focus on urgency and change that is implicit in the Buzan *et al.* concept of securitization even indicates that we are looking for a multiplicity of such statements, an accumulation of statements that represent something as an existential threat to a referent object over a relatively short timespan. The first empirical chapter of this book starts out by reviewing Russian statements on Chechnya during the interwar period broadly and identifying where and how an accumulation of statements on Chechnya as an existential terrorist threat to Russia emerged.

Securitizing narrative

Studying these significative practices in a structured way when working with a given empirical case entails constructing an analytical template outlining the sequence of elements that make up the security argument. Such a template, which I refer to as the *securitizing narrative* throughout the book, denotes the significative aspect of discourse. As noted, discourse is a supra-concept which includes both significative practices (securitizing narrative) and material practices (emergency measures). The securitizing narrative enables us to map out the pattern of argument actually deployed in a given securitization and formalizes how the security argument produces boundaries (between the threat and the referent object) for acceptable action. It captures the distance that is created between Other and Self in representations as well as concomitant policy suggestions on how to treat the Other. Echoing Croft's claim that 'securitization is about shaping relations between identities in particular, and *confrontational* ways',[33] this explication emphasizes that a securitizing 'move' or 'attempt' in the terminology of Copenhagen School securitization theory carries in it a narrative in which Self and Other are represented as radically opposite, so radically opposite that the Other emerges as an *existential* threat to the Self.

In their discussion of the first facilitating condition under which the speech act aimed at securitization works ('the demand internal to the speech act of following the grammar of security'), Buzan *et al.* only hint at how such an analytical template could look when they say that the securitizing discourse is more likely to be authoritative and convincing if it takes the form of a securitizing plot that includes (1) existential threat, (2) a point of no return and (3) a possible way out.[34] Here I will elaborate on how this suggested 'grammar of security' can be explicated as the content of the securitizing narrative.

The first element in the securitizing narrative – existential threat (1) – concerns the description of the nature of the threat and the referent object. In Copenhagen School securitization theory, the notion that the threat is represented as 'existential' is absolutely fundamental, but no tools are offered to establish when a threat representation has reached the level of 'existential'. This maps on to criticism raised by Stritzel on 'the lack of clear criteria for assessing when we have reached beyond the threshold of normality'.[35] Post-structuralist ideas on how identities are constituted suggest the possibility of *scaling* representations of threat.

Post-structuralists understand all social phenomena as being organized according to the same principles as language. Thus, the claims that the structure of language is never totally fixed, and that meaning is constructed through the juxtaposition of signs, have implications for the conceptualization of identity. Identity is conceptualized as relational in the sense that identity is always given through reference to something that it is *not*.[36] In Connolly's sense, identities, whether personal or collective, are not given, but are constituted in relation to difference. Difference is not a given either, but is constituted in relation to identity.[37] Engaging this same conception of identities as mutually constitutive and drawing on Laclau and Mouffe's discourse theory, Hansen suggests that 'meaning and identity are constructed through a series of signs that are linked to each other to constitute relations of sameness as well as through differentiation to another series of juxtaposed signs'.[38] Identities can therefore be said to be highly structured. But again, they are also seen as flexible and changeable entities that can never be completely fixed, because the signs in these chains of sameness and difference may be changed and substituted.

Identities are not necessarily drawn up in relation to radical and threatening Otherness.[39] Yet, given securitization theory's emphasis on existential threat, a post-structuralist securitization theory developed to understand how war becomes acceptable has to focus on radical Otherness.[40] Threat constructions can be placed on a scale with differing degrees of difference and danger attached to them. While some link the issue to descriptors that do not indicate danger or difference in negative terms, other constructions are so radical on these two accounts that the issue emerges as an *existential* threat. In between these two poles, there are threat constructions that indicate varying degrees of danger to and difference from the referent object. While Croft might be right that securitizations hinge on both constructions of Radical Other and Orientalized Other (implying representations of the Other as mystical, attractive and exotic),[41] this second type of Other will be downgraded in securitizations for war and extreme violence. As the case study presented in this book will reveal, orientalizing representations of the Chechens were a part of the Russian discursive fabric on Chechnya historically and before the war, but they all but disappeared in the Russian writ for massive violence during the Second Chechen War.

Reasoning along the same lines as on the scaling of threat representations, the second element within the securitizing narrative – 'point of no return' (2) – can be conceptualized as a scale of alternative futures for the referent object. A future where the referent object cannot exist can then be placed at the top end of this scale, below which there would be possibilities for peaceful co-existence. The third element in the securitizing narrative – 'a possible way out' (3) – identifies the policy or emergency measures necessary, given the gravity of the threat. In this third element of the narrative, we find a description of how to deal with the threat (the policy proposal) in order to achieve a future of survival.[42]

The securitizing narrative is internally consistent. The level of threat implied in the representations (1) delineates a boundary between the threat and the threatened, but also a boundary for acceptable action (3). A threat representation that can be placed at the top of the scale in terms of danger logically fits together with policy proposals that are equally radical or violent. Salter's argument is instructive.[43] He claims that the classification of 'barbarian' is not only dependent on counter-concepts (savage, civilized), but also has effects. The kind of security policy deemed available and legitimate regarding 'barbarians' is other than what is thought of as available and legitimate in other relationships. As suggested above, the link between identity construction and policy option is such that a policy will appear legitimate if it is consistent with the identity construction on which it draws. This means that going *up* the scale of threat representation will legitimize tougher or more violent policies. Put simply, representations of existential threat (i.e. that can be placed at the top end of the scale in terms of difference and danger) make practices of brute violence and war seem logical, legitimate and, maybe, even necessary. Hagmann also alludes to this congruence in the securitization narrative, when he suggests that to name something a 'national danger ... is to stipulate a non-reducible necessity for a national policy response'.[44]

Referent object identity

While the post-structuralist ideas on identity construction and policy making posit that representations of existential threat allow for certain policy proposals, they also, as noted, suggest that changing representations of threat come together with changing representations of the referent object. This insight, that securitization of an issue – identifying something as an existential threat to a referent object – has effects in terms of maintaining and changing the identity of the referent object, has been neglected in Copenhagen School securitization theory.[45] It is particularly relevant when talking about securitization for war. If the threat is described, those who are said to be threatened will necessarily have to be described as well. Given that the changing boundary always has an inside and an outside that are linked, the identity of the referent object will necessarily have to be (re-)defined in relation to the representation of something as an existential threat.

Building on Derrida, the relation constructed in securitizing attempts through a series of juxtapositions between threat and threatened is not neutral in terms of power, as one element (the referent object) will be valued over its opposite (threat).[46] Thus, the re-defining of identity in the face of an existential threat can have substantial effects in terms of cohesion, power and stability within the social entity said to be threatened. With this move, securitization theory can be used to say something more about 'how political communities themselves are constituted', as called for by McDonald.[47] In particular, it can shed light on how securitization for war becomes the key engine in the production of national identity.

Research questions in this book

By putting the practice of securitization at the centre of analysis and operationalizing it as the identification of something as an existential threat to a referent object by way of a securitizing narrative, a first set of questions for the empirical inquiry in this book can be extracted: What identity was 'Chechnya' given in Russian representations (1996–2001)? What level of threat was attached to 'Chechnya' in Russian representations? How has Russian identity been re-drawn in the process of representing Chechnya? How has Russia's future been described? What have been proposed as relevant policies for dealing with Chechnya? These questions will not be addressed in chronological order, but will be investigated within each time period and in different types of texts (see Chapter 3 on sources below). In particular, they will be discussed in Chapter 4, which analyses official and media representations of Chechnya in the interwar period, and in Chapter 5, which analyses official representations of Chechnya and Russia during summer and autumn 1999. They will also re-appear in Chapters 7, 8 and 9, when the official securitizing narrative for war is used to compare representations of Chechnya and Russia in various 'audience' groups during autumn 1999, and in Chapters 11 and 12, when representations on particularly 'shocking events' during the Second Chechen War are reviewed.

Emergency measures

Copenhagen School securitization theory merely states that acceptance of the securitizing argument by the audience leads to endorsement of emergency measures 'beyond rules that otherwise have to be obeyed'.[48] This reads like a positivist set-up of variables that produce an outcome, but gives no clue about what kinds of emergency measures are actually enabled in a process of securitization. Because the Copenhagen School neglects post-structuralist insights, it accepts a divide between significative and material practices and misses the opportunity of theorizing the link between these two in a process of securitization.

In the post-structuralist perspective, meaning and materiality go hand in hand. Building on Foucault, Laclau and Mouffe understand the entire social field as constituted by discursive logic. This means that discourses do more than include systems of signs: they are also material and hence they *encompass* the social field.[49] Put differently, 'discourses are "concrete" in that they produce a material reality in the practices that they invoke'.[50] A post-structuralist framework for the study of securitization can therefore help us theorize the concrete material emergency measures that are enabled in a discourse of existential threat. What I propose is to conceptualize 'emergency measures' in securitization theory as equivalent to the knowledgeable practices that are the material expressions of significative practices and are seen as complementing these in post-structuralist discourse theory.

As Hansen argues, 'while policy discourses construct problems, objects and subjects, they also simultaneously articulate policies to address them. Policies are thus particular directions for actions'.[51] The securitizing narrative presented above captures the 'policy discourse', the significative articulations that contribute to bring something into being as an existential threat, but a full post-structuralist approach to securitization should include the study of how these significative practices find an expression in material practices (emergency measures) and how these material practices in turn serve to constitute and confirm the identity constructions in the securitizing narrative. The radical differentiation between Self and Other in a process of securitization will not only be established in a securitizing narrative. Any group or object that is represented as an existential threat will also be materially constituted as such in a physical space and in the ways it is treated. With this move, the current framework addresses a criticism voiced by Wilkinson that Copenhagen School securitization theory neglects both physical action and how such action contributes to reinforce threat constructions.[52]

As already noted, the changing of representations to foster a relation of radical opposition between Self and an existentially threatening Other would not *determine* emergency action, but would *condition* the range of emergency measures political actors could undertake legitimately. Significative practices condition emergency measures in the sense that they both open up and constrain the range of feasible material practices and actions.[53] Given the internal congruence between identity construction and policy proposal/way out in the securitizing narrative, this explication assumes that certain representations of the Other (such as 'terrorist' or 'infidel') will be followed by policy proposals that permit certain actions (such as killing or torture) while prohibiting others (such as negotiation). To be clear, the assumption is only that the representation ('terrorist', 'infidel') enables the *legitimate* undertaking of a certain type of action (such as killing or torture): this action might still have been undertaken without such a radical representation, but would not have made much sense.

For a book that applies a post-structuralist reading of securitization theory, this means that, in addition to assessing the significative practices through the prism of the securitizing narrative, it is necessary to assess the enactment of this narrative in specific policies and material practices directed towards that/those represented as existential threat. 'Emergency measures' should be studied by investigating the linking of two aspects: the significative representations in the securitizing narrative (particularly the policy proposal/'the way out' given in the securitizing narrative), and the enactment of this in concrete security practices aimed at countering the threat. Then, finally, such a study should assess the way in which these physical security practices contribute to confirm the identity constructions drawn up in the securitizing narrative.

Given the wider conception of the securitization process implied in this re-reading and the emphasis on the 'audience' as contributors to securitization (see 'Audience acceptance and breaking free of rules' below), the material enactment of the narrative will also be carried out beyond the practices of state security agencies. As Croft highlights in his study of how securitization shapes everyday life in Britain, 'the extraordinary measures that are brought forward by a successful securitization are not merely enacted by government, but are the product of wider performances in and throughout society'.[54] This book will reveal how the physical isolation of Chechens in Russian cities by their absence in hired flats, regular jobs and schools, coupled with their presence at police stations and in the zone subjected to bombing have confirmed their identity as different and dangerous in Russian society.

Security practices in rupture, not routine

In order to stay within the bounds of securitization theory and the Copenhagen School conception of security as one implying *urgency, change and extraordinary forms of action*, the study of material practices in this framework will not be directed towards their routinized nature, as is the case in the approach advanced by Didier Bigo[55] and the 'Paris School'.[56] Their suggestion that security evolves in a longer time perspective and is constructed through a range of routinized and often institutionalized practices has opened up a new and fruitful venue in security studies, focusing on 'the creation of networks of professionals of (in)security, the systems of meaning they generate and the productive power of their practices.[57] Yet, with securitization theory as a point of departure, it seems most reasonable to focus on changes in or beginnings of what later become such patterned emergency actions. With its focus on extraordinary means, securitization theory arguably directs our attention more towards how material practices are changed or even established, than to their routinization over time. This is particularly relevant when studying securitization for war: that passing of a threshold which implies that extreme violence is introduced. When something is (suddenly) raised to a level of existential threat, this enables/legitimizes new types of action or – alternatively – intensifies existing security practices.

Securitization as a general mechanism

In order to clarify the conceptualization of emergency measures further and justify its usefulness for understanding the Second Chechen War as an empirical case, yet another question needs to be addressed. What kind of policies and practices are we looking for, what kind of policies and practices qualify as 'emergency measures' in a system that is not liberal–democratic? The undertaking of emergency measures will be understood in the wide fashion actually indicated by Buzan *et al.* In their *Security: A new framework for analysis*, this component is described as undertaking actions beyond 'rules that otherwise have to be obeyed'.[58] This formulation suggests that the legitimate undertaking of 'emergency measures' when something is established as an existential threat is a situation which can occur in *any* political system and society. Securitization is no longer seen as a specific mechanism at work in Western democracies, as implied in the widespread understanding of it as a type of 'special politics' whereby an issue can be moved beyond normal democratic procedures.[59] As Buzan *et al.* point out:

> in other societies there will also be 'rules' as there are in any society, and when a securitizing actor uses a rhetoric of existential threat and thereby takes an issue out of what under those conditions is 'normal politics', we have a case of securitization.[60]

Securitization is a general mechanism that is at work when emergency measures are legitimized in any political system, including when Putin's Russia goes to war. 'Emergency measures' are thus operationalized here as the policies and material practices directed towards Chechnya that were enabled by the establishment of Chechnya as an existential threat and that broke the specific rules of the society and political system of Russia.

Research questions in this book

Based on this conceptualization of emergency measures and how they are legitimized and enabled in a process of securitization, Chapter 4 highlights how the de-securitizing narrative that dominated Russian statements on Chechnya during the inter-war period materialized in concrete cooperative practices. Chapters 10, 11 and 12 review the policies and material practices undertaken as part of the 'emergency measures' for dealing with Chechnya during 1999 and 2000. They investigate how the emergency measures were made logical and legitimate through significative practices in the broader Russian debate during the same period and how they contributed to reinforce and confirm the dominant narrative of Chechnya as an existential terrorist threat to Russia.

Audience acceptance and breaking free of rules

While the current framework broadens the scope of securitization studies to include the material and physical manifestations of a securitizing discourse via post-structuralist insights, it also expands the framework beyond the narrow focus on representations by a 'securitizing actor'. This section first posits that 'actors' are empowered, not before, but *during* a process of securitization, through the productivity of discourse. It then moves on to suggest that 'discursive terrains' as well as representations voiced in the broader public debate, contribute to a situation of 'audience acceptance' and the legitimation of emergency measures.

Actor-hood

If we look at Copenhagen School conceptualizations of speaker and audience and the relation between them, explicated by adding the second and third facilitating conditions in Buzan *et al.*'s book, they can easily be read as contradicting post-structuralist tenets.[61] Turning first to the role of the speaker, Copenhagen School securitization theory emphasizes 'the social conditions regarding the position of authority for the securitizing actor – that is, the relationship between the speaker and the audience and thereby the likelihood of the audience of accepting the claims made in a securitizing attempt'.[62] This suggests that there are limitations as to who can make security claims successfully and that the pre-existing power position of a securitizing actor is important for succeeding with securitization.[63] In line with this emphasis, Copenhagen School securitization theory is, as conceded by Buzan and Weaver themselves 'not dogmatically state-centric in its premises' but 'state-centric in its findings'.[64]

Post-structuralist ideas on the productivity of discourse turn the emphasis on the pre-existing power position of the actor (usually the state) on its head and suggest that securitization of an issue – identifying something as an existential threat to a referent object – has effects in terms of maintaining and changing not only referent object identity but also *actor-hood*. As Stritzel points out, Wæver himself actually opens for such a reading, particularly in his single-authored texts, where he builds on Jacques Derrida's claim that 'there is nothing outside the text' and Judith Butler's idea that speech acts have power to constitute meaning and create new patterns of significance in social relations.[65]

Milliken's description of the *productivity* of discourse highlights this aspect:

> Discourses define subjects authorized to speak and to act ... knowledgeable practices by these subjects towards the objects which the discourse defines,

rendering logical and proper interventions of different kinds, disciplining techniques and practices, and other modes of implementing a discursively constructed analysis. In the process, people may be destroyed as well as disciplined, and social space comes to be organized and controlled, i.e. places and groups are produced as those objects. Finally, of significance for the legitimacy of international practices is that discourses produce as subjects publics (audiences) for authorized actors, and their common sense of the existence and qualities of different phenomena and of how public officials should act for them and in their name.[66]

Drawing on this insight, we can achieve a different conceptualization of actor. In this reading, the authority to speak and act is constituted by the productive power of the discourse itself. It is not inherent in the position of the actor at the outset, but in the process of securitization. When a securitizing argument is launched, it draws up boundaries (by identifying something as an existential threat to a referent object) and limits the range of acceptable policies – thus also producing an actor, by demarcating a sphere in which the actor can then legitimately undertake such policies. According to Jackson:

> a particular deployment always contains one or more *subject-positions* from which action can be taken, and it thus contributes to the production of the actor at the same time as it reveals a particular world in which that actor can subsequently act.[67]

The recognition of existential threat implicit in a securitization creates a particular situation of urgency and thus seems to *require* action by competent actors who, in turn, are empowered to act by this situation. As put by Hagmann, securitization provides both points of reference for agency and *asks for* agency at the same time.[68]

In sum, the re-defining of identity in the face of existential threat can have substantial effects in terms of cohesion, power and stability within the social group that is recognized as referent object, and it also produces an urgent need for an 'actor' and bestows this actor with authority to counter the threat. The question of how the power of an actor can be built through securitization processes will not be addressed in depth in this book, but will be touched upon in Chapter 3 and re-appears in the conclusion (Chapter 13). Anyone studying Russia would agree that the securitization of the Chechen threat contributed greatly to Putin's rising power, although few have tried to find out *how*.

Discursive context and discursive terrains

A post-structuralist securitization theory pays attention to the productive power of securitizing attempts, but it also points to the (con)textual limits such attempts are embedded in at the outset and encounter once they have been launched. As most of the securitization theory post-Copenhagen scholars point out, the Copenhagen

School ultimately downplays the importance of contextual factors.[69] When it does give context some attention, it seems to encourage the analysis of how representations of threat resonate with the external reality. The third facilitating condition 'under which the speech act aimed at securitization works' is suggested by Buzan *et al.* to be 'features of the alleged threats that either facilitate or impede securitization'.[70] As pointed out by Stritzel, this bestows a given external materiality with a causal role beyond its being mediated through language.[71] Such a conceptualization of context cannot be reconciled with the post-structuralist claim that the entire social field is constituted by discursive logic.[72] However, *discursive context* – the discursive structure which securitizing attempts are embedded in and resonate with – should be part of a post-structuralist framework and would allow the theory to move focus beyond the moment of intervention, as called for by McDonald.[73]

The post-structuralist concept of discourse, and more specifically the notion of *intertextuality*, suggest that discursive contexts matter – both historical and specific for different societies and for different ages. Kristeva's (1980) conception of intertextuality implies that all texts are situated within and against other texts: they draw upon them in constructing identities and policies, they appropriate as well as revise them, and build authority by citing them. As Wæver himself notes:

> Discourses organise knowledge systematically, and thus delimit what can be said and what not. The rules determining what makes sense go beyond the purely grammatical into the pragmatic and discursive, linking up to some extent to the traditional studies of 'histories of ideas' in terms of 'how did they think in different periods', or more precisely: how is the conceptual universe structured into which you have to speak when acting politically? Subjects, objects and concepts cannot be seen as existing independent of discourse. Certain categories and arguments that are powerful in one period or at one place can sound non-sensible or absurd at others.[74]

Thus, 'securitizing moves' are not launched into empty discursive space, but into specific cultural contexts. They are structured by and resonate with latent or manifest representations in pre-existing discourses. Any securitizing attempt is launched within a broader discursive context that constitutes it as significant, or not. Existing discourses privilege and disadvantage certain securitizing attempts, as opposed to others.[75]

Within the broader discursive context, several *discursive terrains* can be identified, such as the international discursive terrain or the national discursive terrain. Salter has broken this down even further and investigates the specific terrain of various professions.[76] The national discursive terrain, which is of particular relevance for this study, consists of a plethora of common meanings and identity constructions, among them alternative versions of an issue that is securitized. These meanings and identities have been negotiated over time ((re)produced, confirmed and/or negated in historical, political, media and literary texts) and are specific to the historical and social setting.

The argument of importance to understand how war becomes acceptable is that a securitizing narrative that resonates well with and draws on recurrent common meanings and identity constructions in the national discursive terrain will acquire legitimacy through this resonance. However, that does *not* change the understanding that a securitizing narrative is continuously negotiated, re-phrased or confirmed in representations voiced from the putative 'audience' – to which we return below.

Research questions in this book

The prominent role that discursive context plays in explaining how war becomes acceptable prompts us to ask what the national discursive terrain on 'Chechnya' in Russia looked like prior to 1999. This question will be taken up in Chapter 6, which summarizes representations of Chechnya and Russia in classical Russian literature, as well as in more recent historical and political texts, including those on the first post-Soviet Chechen war. As noted, the national discursive terrain is considered as part of the larger discursive context into which securitizing attempts are launched. Other parts of the discursive context could also be con-sidered in a post-structuralist framework for studying securitization. For example, the re-rephrasing of Chechnya as an international terrorist threat and the acceptance of this re-phrasing by international society cannot be understood without reference to the international discursive terrain at the time. However, that question falls beyond the scope of this empirical inquiry.

Audience acceptance

Turning to the role of the audience in Copenhagen School securitization theory, the emphasis on *intersubjectivity* in the establishment of an existential threat is fully in line with post-structuralist understandings. Buzan *et al.* even make explicit reference to Derrida when they point out that:

> Whether an issue is a security issue is not something individuals decide alone. Securitization is intersubjective and socially constructed: Does a referent object hold general legitimacy as something that *should* survive, which entails that actors can make reference to it, point to something as a threat, *and* thereby get others to follow or at least tolerate actions otherwise not legitimate? This quality is not held in subjective and isolated minds: it is a social quality, a part of a discursive, socially constituted, intersubjective realm (italics in original).[77]

However, one could, as Buzan and colleagues sometimes seem to do, make a leap from this idea of a process of intersubjective establishment of something as an existential threat to a conception of the 'securitizing attempt' as a product of the individual securitizer's words, with the 'audience' as a given entity with a veto role in an attempted securitization and with 'acceptance' as a moment of rational choice. Applications of securitization theory have often treated the audience as a given. In such cases, the audience's preferences will already be fixed, and the audience can reject the threat representation – thus, securitization fails.[78] But such a conceptualization of audience acceptance is at odds with a post-structuralist reading of securitization theory.

Also possible is another reading, one which builds on Wæver's post-structuralist heritage and which is more suitable for this framework. Such a reading entails seeing the audience as a potential field into which the securitizing attempt is launched. Given the malleable yet fixed quality of discourses and the struggles between them, the discursive reception of the securitizing attempt in the 'audience' will be conditioned upon how well it resonates with the discursive terrain, but there is also room for change and appropriation of the securitizing narrative: it can be confirmed, revised – or rejected – by representations in the 'audience'. 'Audience' responses to the securitizing attempt enter the discursive battle on what meaning should be attached to the object. Agreement on something as an existential threat (thereby making possible the legitimate undertaking of emergency measures) is a result of both securitizing attempts and audience responses, and takes the form of a many-layered and dominant discourse.[79] This explication also means that the status of the audience as an audience is ambivalent. The audience is not passive, merely on the receiving end. The audience can also contribute to the securitizing narrative and become part of the 'securitizer'.

The production of the 'consenting audience' becomes a joint act in which both 'securitizing actor' and 'audience' (in the understanding of the Copenhagen School) participate.[80] It should be understood as an intersubjective and negotiated process of legitimation through which sharp boundaries are established between the threat and the threatened as well as the 'way out', ruling 'emergency measures' acceptable. What Buzan *et al.* talk about as 'acceptance of that designation by a significant audience'[81] is, then, in the sense of Laclau and Mouffe, a situation when a particular securitizing narrative has become *hegemonic* in the broader public debate by naturalizing this particular intervention and overpowering others.[82] Empirically, this is the situation when the description of the threat as 'existential' and of 'the point of no return' and 'way out' given in a securitizing narrative (not necessarily promoted from the top of a political system) has gained enough resonance and response in the representations of the 'audience' for emergency action to be undertaken *legitimately.* It is this intersubjective legitimating process that makes it possible to break free of rules that otherwise bind, and undertake emergency measures.

The implication of this reading is that studies of securitizations should not be limited to the statements of state and political leaders, the presumably dominant voices in the construction of security. While most critiques of Copenhagen School securitization on this account have raised the normative problematique implied by the silencing of marginal voices,[83] the current explication critiques the insufficient attention paid to where representations of existential threat actually emerge from and how they become dominant. Security is a site of competing discourses and there are many authoritative utterances beyond those voiced from a political position.[84] This study will uncover how the Russian media contributed heavily to the constitution of Chechnya as an existential threat to Russia, even long before Vladimir Putin came to power, and how it contributed to make this representation absolutely supreme for the Russian public during 1999 and 2000 (Chapter 4 and 9).

Finally, in this reading, securitization is never a stable social arrangement: securitizing claims must be reproduced continually, and no object can become so firmly established as an existential threat necessitating extra-political action that it cannot be challenged. Theoretically, the legitimacy of a policy of war can unravel via a process similar to that which made war acceptable. An intersubjective process which establishes the Other not as an existential threat but as something far less threatening to the referent object would render other policies than war more logical and acceptable. Such a process *can* start from the putative 'audience' with an accumulation of alternative representations of the Other, those representations temporarily outdone by the hegemonic representations of existential threat.[85]

The relevance of audience acceptance

The question of whether securitization presupposes a democratic and rights-oriented political system and consequently that the concept of 'audience acceptance' of security claims is irrelevant in non-democratic political systems has been raised many times.[86] This problematique becomes irrelevant in a post-structuralist reading of securitization theory which takes as its point of departure that *any* policy in any type of political system will rely on intersubjectively constructed accounts that can make these policies appear understandable and legitimate to a potential audience (however small). Buzan *et al.* actually give a similar answer: 'no one is guaranteed the ability to make people accept a claim for necessary security action ... as even communist elites of Eastern Europe learned'.[87] Thus, also in non-democratic systems, leaders legitimize their use of extraordinary measures and these narratives are intersubjectively embedded and constantly subject to some kind of social negotiation in ongoing language use.

Having made the point that 'audience' is significant in any political system, let us turn to the case at hand. In the period between 1996 and 1999, Russia was not a consolidated democracy, nor was it an autocracy. Moreover, the regime was characterized by a presidency that was strong according to the constitution, but in reality quite weak, especially in terms of how contested most of its policies were for the Russian public. This was a situation in which the Russian public's acceptance of security claims articulated by the country's leadership could by no means be taken for granted. Such public endorsement must have seemed highly necessary in order to undertake a new war against Chechnya. In turn, such broad public endorsement in the given Russian situation would prove highly productive in terms of power.

Research questions in this book

This conceptualization of audience acceptance in the process of securitization elicits another line of enquiry to uncover how a new war against Chechnya became acceptable: how was Chechnya as Russia's Other (re)articulated in representations in the wider Russian public debate during autumn 1999? To what extent were representations of 'threat', 'referent object', 'the point of no return', and 'the way out' inherent in the official securitizing narrative negated, rephrased or confirmed in these representations, and how? These questions are addressed in Chapters 7, 8 and 9, which investigate how Russian 'audience acceptance' during autumn 1999 came about. These chapters analyse the significative practices of groups that can be considered key contributors to the wider Russian public debate in 1999: not only the members of the Russian political elite outside of government and situated in key institutional positions such as the Federal Assembly, but also experts and journalists.

Preliminary conclusions on post-structuralist securitization theory and war

The version of securitization theory elaborated in this chapter has been shaped by the urge to understand an empirical case – the broad public acceptance of the Second Chechen War – as well as by a theoretical commitment to a perspective that takes the social constitution of politics seriously in a fundamental way. Key elements of Copenhagen School securitization theory have been amended and new ones developed through post-structuralist insights to make a framework to study how war becomes acceptable.

This has implied conceptualizing securitizing attempts or moves not as speech acts in the Austinian sense but as an accumulation of statements that serve to construct something as an existential threat. I have sought to give content to the definition of the various parts of the securitizing 'narrative' implicit in such a securitizing move, and indicated the possibility of scaling threat in order to determine when a threat can be considered to have reached the level of 'existential'. I have also noted, in line with post-structuralist insights, that the relation between the components in this 'narrative' are of a mutual and co-constitutive nature and that there is a certain congruence between representations of threat and policy proposals. The more radical the difference projected between threat and referent object, the more radical and violent policy suggestions seem logical and legitimate. Moreover, engaging yet another post-structuralist insight neglected in classical securitization theory, the framework has been expanded beyond policy suggestions in a securitizing narrative to include the study of the physical and material practices that befall those classified as existential threat, as well as how the undertaking of such emergency measures confirm and constitute identity-constructions in the narrative. In sum, re-focusing securitization theory back to the 'grammar of security' via post-structuralist insights means that the centre of analysis becomes how securitizing discourse shapes the understanding of the objects of which it speaks and the practices made logical by this understanding.

While stripping the theory of elements such as the pre-existing power position of the securitizing actor and any extra-discursive contextual factors, I have emphasized securitization as a broad, powerful and socially productive process. Securitizations have a history in the sense that they re-build and build on previous, reiterated representations of threat and referent object in a culture-specific discursive terrain. But they are also broad and changeable in real time. What is often construed as the 'audience' contributes to the constant debate on what level of difference and danger to attach to an object. The narrative in a securitizing move can be negated as well as confirmed and expanded on in audience representations. It can even start from voices in the putative audience. Thus, if 'audience acceptance' emerges to legitimize the undertaking of emergency measures against an object or a social group, this is the result of both securitizing moves and audience responses and takes the shape of a many-layered and dominant discourse of existential threat. Such 'audience acceptance' does not give a 'securitizing actor' carte blanche to undertake *any* sort of emergency action, nor is it an instruction to undertake *one* specific emergency action. Rather, the detail in the securitizing narrative (which has been 'agreed' upon) stipulates a *range* of possible and legitimate emergency measures. Within this range there is a certain degree of specificity, in that the level of threat implied in the representation of the object and the suggestions on 'the way out' will indicate what level of violence/force can be employed legitimately.

Copenhagen School securitization theory has been criticized for assuming that construction in the security realm is sufficiently stable in the long run and can therefore be treated as objective.[88] The re-engagement with a meta-theoretical perspective that takes the social constitution of politics seriously shifts attention to how identities are reconstituted and new patterns of significance in social relations are created in a discourse of existential threat. The securitizing discourse does more than form and disempower the object that is said to be threatening. It also empowers the 'referent object' by producing a threatened subject and positioning it 'above' the threatening object, as well as producing a 'securitizing actor' by creating such a 'subject position from which action can be taken'.[89]

This post-structuralist reading of securitization theory indicates several things about securitization for war. First, that broad securitization of something as an existential threat makes possible the legitimate undertaking of war, but also has effects in terms of re-drawing the identity boundary of the group that is said to be threatened and creating unity within this group. Second, that securitizing moves that suggest war as the 'way out' acquire legitimacy if they draw skilfully on ingrained, established and popular representations of threat in a given discursive terrain. Third, if and when war becomes acceptable, this is due to the discursive efforts of 'securitizing actors' and 'audience' alike, because securitization is seen as a broad intersubjective process of legitimation leading up to an agreement on something as an existential threat that necessitates violent reaction. Finally, the type of classification/representation agreed upon during securitization will affect how the war is waged; radical descriptions of the threat legitimize radical means in war. We return to these preliminary suggestions for a post-structuralist reading of securitization and war in the concluding chapter of the book.

Notes

1 The ideas I build on in the framework are extracted mostly from the works of security and identity scholars who explicitly adopt post-structuralism and draw on various post-structuralist thinkers. As Lene Hansen notes, the main approach of post-structuralists in international relations has been to *combine* the positions of, for example, Foucault, Derrida and Butler (L. Hansen. 2011. The politics of securitization and the Mohammad cartoon crises: A post-structuralist perspective. *Security Dialogue*, 42(4/5): 357–369, 358). The framework does not give a general introduction to post-structuralism, but highlights and draws selectively on concepts and ideas relevant for the re-reading of securitization theory.

2 P.T. Jackson. 2011. *The Conduct of Inquiry in International Relations.* New York: Routledge, 115–141. Philosophical wagers are defined as 'philosophical commitments about questions of philosophical ontology that can never be settled definitely' (Jackson 2011, 34–35). Jackson (28 and 34) offers a typology of *philosophical ontologies* understood as 'the conceptual and philosophical basis on which claims about the world are formulated in the first place'. The typology is defined by combinations of two core *wagers*. While the first wager concerns the relation between the knower and the known, the second concerns the relation between knowledge and observation.

3 E. Laclau and C. Mouffe. 1985. *Hegemony and Socialist Strategy: Towards a radical democratic politics.* London: Verso, 108.

4 O. Wæver. 2003. From 'Securitization: taking stock of a research programme in Security Studies.' Unpublished manuscript: 9.

5 B. Buzan, O. Wæver and J. de Wilde. 1998. *Security: A new framework for analysis.* Boulder CO: Lynne Rienner, 5.

6 Buzan *et al.* 1998, 26.

7 Also Floyd and Vouri (2008, 2010) have attempted to stay faithful to the original thoughts of Wæver in their explications of securitization theory (R. Floyd. 2010. *Security and the Environment: Securitization theory and U.S. environmental security policy.* Cambridge: Cambridge University Press; J. Vuori. 2008. Illocutionary logic and strands of securitization: Applying the theory of securitization to the study of non-democratic political orders. *European Journal of International Relations*, 14(1): 65–99; J. Vuori. 2010. How to Do Security with Words: A grammar of securitization in the People's Republic of China. PhD Dissertation at the University of Turku.

8 B. Buzan. 1997. Rethinking security after the Cold War. *Cooperation and Conflict* 32(1): 5–28. Buzan *et al.* (1998, 33) also hold that there are certain facilitating conditions under which the speech act aimed at securitization works: (1) the demand internal to the speech act of following the grammar of security; (2) the social conditions regarding the position of authority for the securitizing actor; and (3) features of the alleged threats that either facilitate or impede securitization. While the first condition concerns the intrinsic features of language and indicates that there is a limitation as to *how* security claims can be made successfully, the other two concern conditions external to discourse. The third seems to indicate that historical and material factors and situations are accorded significance outside their discursive emergence.

9 Buzan *et al.* 1998, 21.

10 Ibid., 4. The 'core insights' of Copenhagen School securitization theory are presented here in a compressed version so that they can function as a point of departure for a post-structuralist re-interpretation. I do recognize, however, that what the key authors of securitization theory actually deliver is not such a monolithic approach.

11 Buzan *et al.* 1998: 26, see also R. Taureck. 2006. Securitization theory and securitization studies. *Journal of International Relations and Development*, 9(1): 52–61.

12 D. Campbell. 1992. *Writing Security*. Minneapolis, MN: University of Minnesota Press; L. Hansen. 2006. *Security as Practice*. New York: Routledge.

13 Hansen 2006, Preface/xvi.

14 Campbell 1992, 4.

15 Post-structuralists adopt a non-causal epistemology, and claim that identity cannot be defined as a variable that is causally separate from foreign policy. One cannot measure identity's explanatory value in comparison to material factors because material factors and ideas are intertwined to such an extent that the two cannot be separated from each other. They are mutually constitutive and discursively linked (Hansen 2006, 25–28).

16 P. Jackson. 2006. *Civilizing the Enemy: German reconstruction and the invention of the West*. Ann Arbor, MI: University of Michigan Press, 16.

17 J.L. Austin. 1962. *How to Do Things with Words*. Oxford: Clarendon Press.

18 O. Wæver. 1995a. Securitization and desecuritization. In: R.D. Lipschutz (ed.) *On Security*. New York: Columbia University Press, 55.

19 For a similar reasoning on why speech act theory is not so useful in studies of securitization, see S. Guzzini. 2011. Securitization as a causal mechanism. *Security Dialogue*, 42(4/5): 329–341, 335.

20 See, for example, B. Buzan and O. Wæver 2003. *Regions and Powers*. Cambridge: Cambridge University Press, 491.

21 See also T. Balzacq. 2005. The three faces of securitization: Political agency, audience, and context. *European Journal of International Relations*, 11(2): 171–201, 182–183; Taureck 2006, 52–61; H. Stritzel. 2014. *Security in Translation: Securitization theory and the localization of threat*. London: Palgrave Macmillan, 20–24.

22 For a post-structuralist view of the agent–structure debate, see R. Doty. 1997. Aporia: A critical exploration of the agent–structure problematique in international relations theory. *European Journal of International Relations*, 3(3): 365–392.

23 For an instructive discussion on Saussure's impact on discourse theories, see M. Jorgensen and L. Phillips. 2002. *Discourse Analysis as Theory and Method*. London: Sage.

24 J. Derrida. 1981. *Positions*. Chicago, IL: University of Chicago Press.

25 J. Milliken. 1999. The study of discourse in international relations: A critique of research and methods. *European Journal of International Relations*, 5(2): 225–254, 230.

26 This elaboration is taken from Jørgensen and Phillips's (2002, 24–59) introduction to Laclau and Mouffe's Discourse Theory.

27 Judith Butler's notion of performativity is defined as 'the reiterative power of discourse to produce the phenomena that it regulates and constrains' (J. Butler. 1993. *Bodies that Matter: On the discursive limits of sex*. London: Routledge, 2).

28 R.R. Krebs and P.T. Jackson. 2007. Twisting tongues and twisting arms: The power of political rhetoric. *European Journal of International Relations*, 13(1): 35–66, 45.

29 Ibid., 45.

30 Buzan *et al.* 1998, 25.

31 Ibid., 32

32 As several scholars have pointed out, meaning is not only communicated through language, but images and visual representations should also be considered as vehicles for securitization. See M.C. Williams. 2003. Words, images, enemies: Securitization and international politics. *International Studies Quarterly*, 47(4): 511–531; F. Møller 2007. 'Photographic interventions in post-9/11 Security Policy'. *Security Dialogue*, 38(2): 179–196; Hansen 2011.

33 S. Croft. 2012. *Securitizing Islam: Identity and the search for security.* Cambridge: Cambridge University Press, 15, emphasis added.

34 Buzan *et al.* 1998, 33.

35 Stritzel 2014, 35.

36 This is an old theme. However, the breakthrough for a method to grapple with this theme came with Fredrik Barth (F. Barth, (ed.) 1969. *Ethnic Groups and Boundaries.* Oslo: Norwegian University Press).

37 W.E. Connolly. 1991. *Identity/Difference: Democratic negotiations of political paradox.* Ithaca, NY: Cornell University Press.

38 Hansen 2006, 42; Laclau and Mouffe refer to this as 'the logic of equivalence' and 'the logic of difference'; see Jørgensen and Phillips (2002, 43–47).

39 Much post-structuralist work within IR after the Cold War has been devoted to exploring how identities have been constructed in relation to other forms of otherness than radical otherness. Wæver (1996) for example has argued that the EU's constitutive Other was its own past, whereas other scholars have explored competing constructions and ambiguities within state identities (I.B. Neumann. 1996. *Russia and the Idea of Europe.* London: Routledge on Russian identity). Hansen (2006) outlines spatially, temporally and ethically constituted identities and Croft (2012, 91) presents a whole range of different forms of Otherness.

40 Connolly 1991; Campbell 1992.

41 Croft 2012, 89–90.

42 This grammatical structure of the securitizing narrative is in some ways similar to the one suggested by Vouri (2008) and expanded on by Stritzel (2014, 48–49) as 'a sequence of claim, warning and demand ... supported by the propositional content.' However, as the current explication draws on post-structuralist insights and ideas of how identities are constructed, it provides an additional logical linking of the different components in the narrative.

43 M.B. Salter. 2002. *Barbarians and Civilization in International Relations.* London: Pluto Press.

44 J. Hagmann. 2015. *(In-) Security and the Production of International Relations: The politics of securitization in Europe.* London: Routledge, 7.

45 As Huysmans (J. Huysmans. 1998. Revisiting Copenhagen: Or, on the creative development of a security studies agenda in Europe. *European Journal of International Relations*, 4(4): 479–505, 489) notes, *Identity, Migration and the New Security Agenda in Europe* (Wæver *et al.* 1993) did introduce the question of how threat definitions have an impact on the identification or constitution of society, but this understanding was bracketed in their presentation of European identity in the book itself. Moreover, it has not been expanded on in later works from the Copenhagen School.

46 Derrida 1981..

47 M. McDonald. 2008. Securitization and the construction of security. *European Journal of International Relations*, 14(4): 563–587, 565; see also critique by Hagmann 2015, 27.

48 Buzan *et al.* 1998, 25.

49 Laclau and Mouffe 1985, 108.

50 C. Hardy, B. Harley and N. Phillips. 2004. Discourse analysis and content analysis: Two solitudes. *Qualitative Methods: Newsletter of the American Political Science Association Organized Section on Qualitative Methods*, 2(1): 19–22, 20.

51 Hansen 2006, 21.

52 C. Wilkinson. 2007. The Copenhagen School on tour in Kyrgyzstan. *Security Dialogue* 38(1): 5–25.

53 With this elaboration of how representations condition policies, the current approach to securitization aligns with that of Hagmann (2015, 8–9), but departs from it in that I do not invoke this relation as one of *causation.*

54 Croft 2012, 84.

55 D. Bigo. 2002. Security and immigration: Towards a critique of the Governmentality of unease. *Alternatives*, 27 (1): 63–92.

56 C.A.S.E. Collective. 2006. Critical approaches to security in Europe. A networked manifesto. *Security Dialogue*, 37(4): 443–487.

57 Ibid., 458.

58 Buzan *et al.*'s full formulation is 'When does an argument with this particular rhetorical and semiotic structure achieve sufficient effect to make an audience tolerate violations of rules that would otherwise have been obeyed?' (1998, 25). They phrase this notion of how audience sanctioned securitizing talk enables the legitimate violations of rules in several different ways. On page 31, they talk about this as a situation in which the audience will 'tolerate actions otherwise not legitimate'; on page 24, they say that 'the issue is represented as an existential threat, requiring emergency measures and justifying actions outside the normal bounds of political procedure'.

59 Huysmans 1998; Williams 2003.

60 Buzan *et al.* 1998, 24.

61 Ibid., 33.

62 Ibid.

63 In contrast to the framework forwarded in this book, which focuses on the productive power of discourse, Stritzel (2014, 29) has elaborated on this line of reasoning which builds on Bourdieu's suggestion that the power of the speech act springs from *institutional power*, i.e. that the authority to speak is vested in the institutional position of the speaker.

64 Buzan and Wæver 2003, 71.

65 H. Stritzel. 2007. Towards a theory of securitization: Copenhagen and beyond. *European Journal of International Relations*, 13(3): 357–383, 361–362; see also Stritzel 2014, 24–27.

66 Milliken 1999, 229, emphasis in original.

67 Jackson 2006, 30.

68 Hagmann 2015, 22, emphasis added.

69 McDonald 2008, 571; T. Balzacq (ed.). 2011. *Securitization Theory: How security problems emerge and dissolve*. Abingdon: Routledge; Stritzel 2014; Hagmann 2015, 23–24.

70 Buzan *et al.* 1998, 33.

71 Stritzel 2014, 33.

72 Laclau and Mouffe 1985, 108.

73 McDonald 2008, 564.

74 O. Wæver. 2002. Identity, communities and foreign policy: Discourse analysis as foreign policy theory. In: L. Hansen and O. Wæver (eds.) *European Integration and National Identity*. London: Routledge, 29.

75 For a similar argument, see Stritzel 2014, 46.

76 M.B. Salter. 2008. Securitization and desecuritization: Dramaturgical analysis and the Canadian Aviation Transport Security Authority. *Journal of International Relations and Development*, 11(4): 321–349.

77 Buzan *et al.* 1998. 31.

78 See, for example, C. Wagnsson. 2000. *Russian Political Language and Public Opinion on the West, NATO and Chechnya*. Stockholm: Akademitryck AB Edsbruk.

79 This conceptualization is in many ways similar to what McDonald (2008, 572) hints at when referring to Judith Butler's suggestion that, if one builds on the speech act in its 'perlocutionary' (necessary for enabling particular actions) and not 'illocutionary' form, this would 'enable greater attention to audiences who might either consent to particular actions suggested through speech or engage in contesting the terms of the speech act or the actions suggested in response to it.'

80 As Patrick Jackson notes, public legitimation cannot be firmly segmented into a

moment of transmission and a moment of reception: it is transactive all the way down (email exchanges between Patrick Jackson and the author in January 2009).

81 Buzan *et al.* 1998, 27.

82 J. Torfing. 1999. *New Theories of Discourse: Laclau, Mouffe and Zizek.* Oxford: Blackwell, 103.

83 For an overview of critiques, see McDonald 2008, 573–575; also L. Hansen. 2000. The little mermaid's silent security dilemma and the absence of gender in the Copenhagen School. *Millennium*, 29(2): 289–306.

84 For a similar critique, see R. Doty. 1998–1999. Immigration and the politics of security. *Security Studies*, 8(2–3), 71–93; and Croft 2012, 81–88.

85 Albeit framed as a language game, Donnelly's securitization framework has similar emphasis on change and multiplicity in the speaker–audience relationship and how security is constantly ascribed with meaning at different stages of the securitization process (F. Donnelly. 2013. *Securitization and the Iraq War: The rules of engagement.* New York: Routledge, 52 and chapter 3).

86 C. Aradau. 2004. Security and the democratic scene: Desecuritization and emancipation. *Journal of International Relations and Development*, 7(4): 388–413; Wilkinson 2007; Vuori 2008; K. Åtland and K. Ven Bruusgaard. 2009. When security speech acts misfire: Russia and the Elektron incident, *Security Dialogue*, 40(3): 333–353; J. Hayes. 2009. Identity and securitization in the democratic peace: The United States and the divergence of response to India and Iran's nuclear programs. *International Studies Quarterly*, 53(4): 977–999.

87 Buzan *et al.* 1998, 31; Juha Vuori makes a related argument in his study of securitization in the Chinese political system, pointing out that 'legitimacy is perhaps the most significant element in the survival of any social institution and all governments must exercise a minimum of both persuasion and coercion in order to survive' (Vuori 2008, 68).

88 B. McSweeney. 1996. Identity and security: Buzan and the Copenhagen School. *Review of International Studies*, 22(1): 81–93; D. Mutimer. 2007. Critical security studies: A schismatic history. In: A. Collins (ed.) *Contemporary Security Studies.* Oxford: Oxford University Press, 53–75. McSweeney was the first to raise this criticism in Identity and security: Buzan and the Copenhagen School (1996), to which Buzan and Wæver replied in Slippery? Contradictory? Sociologically untenable? The Copenhagen School Replies, (1997).

89 Jackson 2006, 30.

3 Method and sources

The choice of research method has been dictated by the research questions and the theory framework which structure this study. The intention is not to explain *why* Russia and Chechnya were at war, but to understand *how* going to war was made possible by representing Chechnya as an existential threat, and *how* shifting representations of Chechnya made certain practices of war possible while precluding others. Moreover, the epistemological and ontological underpinnings of securitization theory adopted with a post-structuralist bias render some version of discourse analysis not only suitable but indeed necessary. If discourse itself is seen to be constitutive of social reality, then findings from investigations at the level of discourse should be significant. Investigation at the level of discourse implies using the text for what it is, not as a sign of something else. The aim is not to try to get behind the text, seeking to find out what actors *really* think and mean when they say or do this or that. I am interested in how people's sayings and doings shape and change our social world.

In the present study, the use of discourse analysis has entailed investigating Russian texts to ascertain how the boundaries of 'Chechnya' as well as the boundaries of Russian identity have been (re-)drawn over time, *and* identifying how policies and material practices with regard to Chechnya have changed with shifting representations. Based on the understanding that collective identities are constructed in processes of linking and differentiation,[1] the texts have been analysed by taking 'Chechnya' and 'Russia' as 'nodal points' and investigating how 'Chechnya' and 'Russia' have been filled with meaning relationally by being equated with some signifiers and contrasted with others.[2] This has been done by reviewing *explicit articulations* of key representations of identity in the texts.[3] For example, 'Chechnya' might be equated with 'culprit', 'criminal' or 'anarchy', while simultaneously differentiated from signifiers such as 'victim', 'law', 'order' and 'civilized' (equated with 'Russia' as referent object).

Jennifer Milliken talks about this as 'predicate analysis', which focuses on the verbs, adverbs and adjectives that attach to nouns.

> A set of predicate constructs in a text defines *a space of objects* differenti-
> ated from, while being related to, one another ... Predicate analysis involve
> drawing up lists of predications, attaching to the subjects the text construct

and clarifying how these subjects are distinguished from and related to one another.

Moreover, the object spaces identified in the *different texts* should be compared to 'uncover the relational distinctions that arguably order the ensemble, serving as a frame (most often hierarchical) for defining certain subject identities'.[4] Since this is a study of securitization – which implies that something/the object is increasingly identified as a *threat* – the 'securitizing narrative' and the components in this narrative ('existential threat', 'point of no return' and 'way out') have been used as an analytical template through which to study representations and determine the detail of a discourse. I have made lists of predications attached to the subjects and compared them over time, looking for a possible *escalation of danger* in representations of 'Chechnya' or other 'events within events' (see below). The level of threat in a representation has been determined by investigating the predications and how they are combined in the statements. For example, signs beside 'terrorist' that are linked to 'Maskhadov' are significant. A representation that couples 'terrorist' to a further construction of 'Maskhadov' as 'non-human' and 'incapable of change' will indicate a higher level of threat against Russia than one that couples 'terrorist' to a further construction of 'Maskhadov' as 'moderate' and 'captive of the radical forces'. While the first construction could indicate a policy of assassination as a possible 'way out', the second would provide an opening for a policy of cooperation and negotiation.

The representations of 'Chechnya' and 'Russia' read through the template of the securitizing narrative have been investigated in a series of texts, both parallel in time and over time, to reveal the relational distinctions drawn up in several discourses and how these change and are contested over time. The focus has been on discovering how, over time, different discourses in Russia have sought to fill 'Chechnya' with various types of content by equating 'Chechnya' to different signifiers. This mapping of representations has revealed the discursive struggles over the kind of security challenge 'Chechnya' *is*, and the types of policies that are suitable for dealing with Chechnya.

As noted by Hansen, policy debates – such as the debate on Chechnya evolving in Russia – are usually bound together by a smaller number of discourses. It is useful to identify some 'basic discourses' in order to identify a possible struggle between them or reveal challenges to an otherwise hegemonic discourse. In this book, I have identified two or three basic discourses within the Russian debate which place 'Chechnya' differently on the scale of threat, suggesting different policies on Chechnya (e.g. 'discourse on reconciliation' and 'discourse on war'). I also consider whether one such discourse acquired hegemony, and whether this hegemony was challenged by other discourses over the timespan covered here. This also enables me to identify 'securitizing (and de-securitizing) actors' in the sense of the Copenhagen School throughout the period under study, although that is not a main focus.

This study focuses on one *event*: Russian securitization of the Chechen threat; but I also investigate the discursive constructions of 'events within events' in the

Russian debate on Chechnya.[5] Examples of such 'events within events' in the period before 2001 include the peace deal that ended the First Chechen War (the Khasavyurt Accord), the interwar domestic situation in Chechnya and the Ichkerian President Aslan Maskhadov. Studying 'Maskhadov' as an 'event within events' has meant taking this sign as the 'nodal point' in the discourse analysis and looking for the predicates attached to it. Studying the changing representations of these smaller but related events serves the purpose of checking, validating and underscoring the findings on the *core event* – 'Chechnya'.

In practical terms, the mapping of representations during work on this book has often entailed constructing charts and placing statements and representations that are similar under the heading of a certain basic discourse broken down to 'Chechen Other/level of threat', 'point of no return', 'Russian Self' and 'policy recommendation/way out'. I include reference to many quotes, but not all: sometimes I have registered a statement simply by ticking the boxes of a certain basic discourse in a chart to show that such representations have been repeated. Through such meticulous registration it has been possible to measure how strong or 'thick' (alternatively, how weak) a certain basic discourse has been.

Obviously, there is and should be a strong quantitative element to discourse analytical work. A discourse is not a statement: it is a thick grid of hundreds of statements that shape social reality. Too often discourse analytical studies make claims on weak grounds, by merely mentioning a few quotes to illustrate what is then held to be a dominant discourse. A few quotes are not enough to substantiate the existence of a dominant discourse. It is necessary to investigate statement after statement, to register detail and changes in representations, as well as to detect the weaker yet emerging discourses of the future. Thus, a guiding principle throughout the work on this book has been to ensure that the number of statements reviewed and charted is high enough to substantiate and validate the claims I make about the shifting patterns of meaning attached to 'Chechnya' in Russia. Finally, in line with the understanding of discourse as encompassing the social field, discourse analytical investigations such as this one should include systematic charting of material practices enabled by significative practices. The final empirical chapters (as well as Chapter 4, to some extent) move beyond using the sources to study significative practices to studying the material practices undertaken during the Second Chechen War.

Intertextual scope, sources and operationalization chapter by chapter

The scope of this study is not limited to official political texts but includes the study of political elite, journalistic, expert, military, security and to some extent classical literary texts and how they interact. Such a model can capture how official discourse is fed, reproduced or contested across a range of sites and how the 'discourse of war' comes to dominate a larger public. The selection of texts is partly directed toward revealing where the 'discourse of war' emerged, but primarily how it was received, revised and confirmed in the texts of the putative 'audience' after being launched from the official political level, and finally how it was enacted in material practices. While the scope of texts has been broadened to include texts beyond the formally political, the intertextual scope is still limited. Popular fiction (e.g. Russian television series and popular literary fiction) is *not* assessed. That is not to say that such texts have not made an imprint on the discourses on Chechnya in Russia or contributed to legitimizing violent practices. They certainly have – but investigating them lies beyond the practical scope of this study.

Concerning the selection of texts within the scope decided upon in the inter-textual model, Hansen proposes three criteria: they should be characterized by the clear articulation of identities and policies; they should be widely read and heeded; and they should have the formal authority to define a political position.[6] Some texts used in this study, such as the statements of Prime Minister/President Putin, meet all these criteria. Statements by the President, other top officials or members of the Russian Federal Assembly quoted in the press, and particularly those transmitted via television, also meet all three criteria. Other texts, such as Duma or Federation Council debates, meet the first and the third, but not neces-sarily the second criterion. The journalistic accounts and opinion pieces by experts reviewed in this book meet the first two criteria, but lack the formal authority to define a political position. Nevertheless, given the topic under study – how war becomes acceptable – and the intersubjective nature of such a social process, the authority and power of expert and journalistic texts seem to justify their centrality to this study. I also rely on a few even more marginal texts, such as texts from the security services, the military and classical Russian literature on the Caucasus. This has been important in order to reveal where discourses emerged or where resistance or re-articulations might emerge in the future, as well as to indicate how far down in Russian society the 'discourse of war' has penetrated.

Two Russian newspapers dominate the source-basis of the book: *Nezavisi-maya Gazeta (NeGa)* and *Rossiyskaya Gazeta (RoGa)*. *NeGa* is a large-circulation, influential newspaper that carried extensive, detailed and many-sided reporting on Chechnya during the First Chechen War. It is also the newspaper that offered the most extensive coverage of the violent conflicts in places such as Nagorno Karabakh, Pridniestr and South Ossetia in the early 1990s. Like almost all Russian newspapers, *NeGa* did not send its own journalists to Chechnya during the Second Chechen War. However, the newspaper sought to maintain an independent position. One indication was the publication of an interview in *NeGa* with the Ichkerian President Aslan Maskhadov in February 2000, despite the prohibition against printing interviews with members of the armed resistance. *Rossiyskaya Gazeta (RoGa)* was chosen because it has always been a mouth-piece of the state, presenting official positions and statements, as well formal official documents such as laws and decrees. The general strategy in work on this book has been to follow every single issue of these two newspapers over a long time-span: 1996–2000. While such day-to-day reading of *NeGa* and *RoGa* has made up the core source of analysis for many chapters, I also conducted searches through the database Public.Ru (which covers thousands of articles from nearly all Russian newspapers) in order to sample articles from other large, mainstream newspapers, and check and adjust the general patterns of discourse found in *NeGa* and *RoGa* articles.

Chapter 4 builds on an extensive body of general material to provide a basis for quantitative identification of the basic discourses on Chechnya in Russia. I consulted the entire volumes of *NeGa* from August 1996 until August 1999 and also the archives of Radio Free Europe/Radio Liberty (*RFE/RL*), tracing all statements on Chechnya by Russian officials and politicians referred to there. These were supplemented by articles on key events in Chechnya (in the interwar period) from other Russian newspapers retrieved through the database Public. Ru. All the *NeGa* 'field reports' on Chechnya for this period created the basis for drawing conclusions on the media discourse on Chechnya in the interwar period. Casting the net beyond the texts of top officials also enabled me to identify the basic discourses in the wider Russian debate, including the more marginal but upcoming discourses and where they emerged in the interwar period.

The key aim in *Chapter 5* has been to uncover the discourse through which Russian state action on Chechnya was legitimized. I have investigated texts from political leaders with official authority concerning Russian policies on Chechnya (the President, the heads of the Presidential Administration, members of the Security Council, the Prime Minister, the Minister for Internal Affairs, the Minister for Foreign Affairs, the Defence Minister), as well as from those with central roles in executing these policies, such as high-ranking military and security staff and senior civil servants. The texts for Chapter 5 thus include interviews and speeches, as well as media reports from Duma and Federation Council debates and statements by officials (referred to in official newspapers such as *RoGa* and more independent ones such as *NeGa*, as well as in other Russian newspapers searched through Public.Ru). *Chapter 6* presents an outline of historical representations of 'Chechnya' and 'Russia', as read through the template of the securitizing narrative, and then compares these with the official securitizing narrative extracted in Chapter 5. This has enabled me to evaluate the 'discursive terrain' already existing in the Russian audience and how well the official securitizing narrative resonated with this. The sources used in this chapter are mostly secondary.

The representations of 'Chechnya' and 'Russia' offered by members and potential members of the Russian Federal Assembly (referred to as the political elite), experts and journalists during autumn 1999 are presented in *Chapters 7, 8* and *9*. The analysis was conducted in two phases. First, in each of these groups of texts, I mapped the identities and policies articulated, using the template of the securitizing narrative, and determined the struggles between basic discourses. And second, I compared the representations in each of the groups of texts with those that made up the official securitizing narrative (extracted in Chapter 5). The overall aim has been to reveal how the intersubjective process unfolded, by investigating similarities, differences and changes in representations in and across the different groups of texts, and in particular in relation to the official texts. In this way, I have been able to establish how far the process of producing a consenting audience evolved during autumn 1999 and how this new public consensus on the necessity of using violence against Chechnya came about.

The sources used to investigate representations among the political elite, the experts and the media during autumn 1999 are primary. Political elite texts are taken from media accounts that directly refer to statements or speeches by members of the Russian Federal Assembly. Newspaper opinion pieces (generally from *NeGa* and *RoGa*) make up the body of expert texts. In all, more than 30 opinion pieces have been investigated, 21 of which are referred to in this book. Concerning journalistic texts, hundreds of newspaper articles (field reports, portraits, chronicles of events and a few editorials) make up the body of texts I investigated in order to draw conclusions on media discourse. In some cases, the use of pictures and the placing of headlines and articles are assessed as well. Such material has not been used as a source of 'facts': it has helped me pin down the kinds of meaning given to 'Chechnya' and 'Russia' in Russian newspaper accounts in the course of autumn 1999, how these representations interplayed with official representations and whether they served to legitimize a policy of war.

Chapters 10, 11 and *12* turn from the dominant, significative patterns on 'Chechnya' and 'Russia' to the question of emergency measures: the policies and material practices that are legitimized and enacted in a discourse of existential threat. This poses some challenges in terms of methodology, because I move beyond using my sources to identify significative practices and into using newspaper articles and human rights reports as sources to establish material practices. How is it possible to establish what the material practices in Russian relations with Chechnya looked like in autumn 1999 and 2000? With all the disinformation, psyops (psychological operations) and counter-information surrounding events in Russia (and particularly those relating to Chechnya), many people have long since given up trying to establish what *actually* happened. Nevertheless, in order to address the entire process of securitization and discourse in full, I have had to venture into trying to establish what material practices on Chechnya amounted to during 1999–2000. I have chosen to rely on human rights reports, to some extent official information, legal documents, Russian and English news reports, as well as secondary accounts that are well researched. I do not claim that the outline of practices presented on the basis of this body of sources is the full and true story of what happened. My intention has been to collect enough data to substantiate the claim that certain material practices existed during these years in Russia. The point is, as noted, not to establish exactly what such practices amounted to per se, but how they were enabled and legitimized by linguistic representations. To establish this link, Chapters 10, 11 and 12 repeat many of the findings on linguistic representations from Chapters 5, 6, 7, 8 and 9, while also building on the study of new texts. These new texts, by military and security personnel of different ranks, have been investigated by means of the same technique as before: reading texts through the prism of the securitizing narrative. With the inclusion of texts from security practitioners on the ground, the book spans both the macro- and micro-level of the Russian debate on 'Chechnya,' from president to foot soldier.

Finally, building on the claim that securitization is never a stable social arrangement and the possibility of change that discourse theory assumes, Chapters 11 and 12 also examine how potentially 'shocking events' (such as gross violations of human rights, or the killing of civilians) during the Second Chechen War were represented in official texts, as well as in political elite, expert and journalistic texts. I selected specific events, such as the bombing of the villages of Elistanzhi, Samashki and Novy Sharoy, of the Grozny market and of a Red Cross-marked civilian convoy, and then reviewed statements on these events in *NeGa* and *RoGa* as well as in 50 newspaper articles retrieved through Public.Ru. Similarly, the *zachistki* (cleansing operations) of the villages of Alkhan-Yurt, Staropromyslovsky, Novye Aldy, Sernovodsk and Assinovskaya and the 'filtration point' at Chernokozovo were selected as potentially 'shocking events'. Statements and reports on these events referred in *NeGa*, *RoGa* as well as in 50 other newspaper articles retrieved through Public.Ru were studied in order to reveal changes in representations of 'Chechnya' and 'Russia' in official language as well as in that of the political elite, experts and journalists. Once again, the securitizing narrative functioned as the core template through which to read the texts.

On translations, referencing and transliteration

This book presents many quotes originally written in the Russian language; I have translated most of them myself. Statements marked with ' ' throughout the book are direct citations given in Russian newspapers. Statements given without ' ' are taken from Russian newspapers but have not been indicated as direct references in the article. In cases where statements have been taken from English-language newspapers or academic accounts, I have relied on their translation from Russian and their indication of whether the statement is a direct quote or not.

The newspaper articles used are mostly referred to only with their Russian or English title, name of newspaper and date.[7] This I have done when referring to a quote (by a politician, official etc.), and is logical, since the focus is on the statement itself and not on the journalist referring to the statement. Also when media discourse is analysed, the names of the journalists are not presented, as I have wanted to direct the focus away from individual journalists, in order to grasp the general broad movement of discourse across the journalist corps. This approach is also appropriate given the limited attention devoted to individual agency in discourse analysis. However, opinion pieces by experts are presented with name, academic degree and (sometimes) affiliation, as I consider this background information relevant for understanding the context of expert discourse.

The system of transliteration used throughout this book is the BGN/PCGN 1947 System. However, I have employed a simplified standard for names of persons and places: -y for -ий -ый endings (and not -iy or -yy), and soft or hard signs are not indicated (e.g. Yeltsin, not Yel'tsin). In general -ц and -тс are both transliterated -ts (not -ts and -t•s). Discrepancies with the BGN/PCGN 1947 System have been allowed in direct quotes when the author of the given text uses a different style of writing Russian names/places, and with works/articles written by Russian authors whose name(s) in the work/article in question have been written according to a different standard.

Notes

1 L. Hansen. 2006. *Security as Practice*. New York: Routledge, 42.
2 Laclau and Mouffe theorize a discourse as formed by the partial fixation of meaning around certain nodal points, a privileged sign around which other signs are ordered. The nodal point in itself is empty, so there is always the possibility of contestation as to what meaning this sign should be invested with (M. Jørgensen and L. Phillips. 2002. *Discourse Analysis as Theory and Method*. London: Sage, 26–28).
3 See Hansen 2006, 53.
4 J. Milliken. 1999. The study of discourse in international relations: A critique of research and methods. *European Journal of International Relations*, 5(2): 225–254, 232–233.
5 On events see Hansen 2006, 80.
6 Hansen 2006, 85.
7 It is a simple matter to locate a particular article in a newspaper base if one has the full title of the article; the author's name is not necessary. The *RFE/RL Newsline* articles are not referred to by title, but only by date and can easily be found in RFE/RL archives by searching for the date, (available at www.rferl.org/search/?k=newsline%20 archive#article and accessed 5 November 2013). The title is referred to for longer RFE/ RL items such as *RFE/RL Features*.

4   The interwar period: a case of de-securitization

Beginning this study of the Second Chechen War with a detour back to official Russian discourse on Chechnya in the interwar period serves several purposes. First, it shows how much Russian representations of Chechnya can change. Historical scholarly accounts, written as well as oral, tend to emphasize Russia's negative representations of Chechens and its harsh and brutal approaches towards them. In all the texts and talks on the subject of Chechen–Russian relations I have read and heard over the years, the words of General Yermolov – 'there is no people under the sun more vile and deceitful than this one' – must be one of the most quoted, along with Lermontov's 'Cossack lullaby' featuring the 'wicked Chechen' who 'whets his dagger keen'.[1] Brutal Russian warfare in the Caucasus in the nineteenth century and the 1944 deportation of the entire Chechen population to Central Asia are core features of any historical account of Russia's encounter with the Chechens. Today, after years of hostile Russian language and policies resulting in war and destruction, it is difficult to imagine a different Russian approach. Re-visiting the official Russian discourse on Chechnya in the interwar period can provide a reminder that, even if Chechnya is one of Russia's habitual Others, it has not always been represented in terms of radical and dangerous Otherness. Moreover, Russia's approach to Chechnya is not doomed to repeat itself forever, nor has it always remained the same. While there is clearly continuity, there is also change.

This first empirical chapter explores linguistic patterns *in and across* official Russian statements by using the 'de/securitizing narrative' and its details ('existential threat', 'point of no return' and 'way out') as a template for eliciting the content of many statements over time. This is in line with the understanding of a securitizing or de-securitizing move as an 'accumulation' process that emerges when *many* statements combined represent an object as an existential threat – or when many statements combined represent an object as non-threatening and close to the referent object. On the basis of such an exploration of official statements in the interwar period, it is possible to identify two or more *basic discourses* on Chechnya, and to determine whether they serve to securitize or de-securitize 'Chechnya' as an object, as well as which of them are dominant.

Apart from this broader objective of documenting changing official representations of Chechnya, I seek here to understand the absence of war between Russia and Chechnya in the years 1996 to 1999 – the 'interwar years'. The proposition offered by a post-structuralist version of securitization theory is that a discourse downplaying Chechnya as a threat dominated the Russian official debate in this period, making other policies toward the republic more legitimate and possible than those requiring the use of force. We should note that there was a period of 'war fatigue' after the conclusion of the First Chechen War in 1996. It could be argued that Russian leaders had no choice but to moderate their enemy image of Chechnya: after all, they had lost the war and had been forced to negotiate a peace deal – elevating Chechnya as a security dilemma was, in a sense, a course of action simply not possible then. However, as time passed and Chechnya slid into de facto independence and chaos, one could well have imagined a new Russian campaign – but this did not happen.

In this chapter, I present two basic official discourses on Chechnya, the struggles between them and how they contributed in shaping Russian policies on Chechnya. The argument which drives this chapter is that, although competing positions on 'Chechnya' did exist, a de-securitizing discourse of reconciliation dominated official statements in this period, rendering impossible a policy of war against the republic. While the main focus is on linguistic practices, I also comment on the policies and material practices undertaken against Chechnya, enabled by the emergence of a de-securitizing discourse at the official level. The chapter also includes a rough outline of what the Russian press reported from Chechnya in this period and how these representations came to feed into the discursive struggle. The account will not include representations of Chechnya in the wider public debate of the time: what I present and evaluate here are de/securitizing moves and practices by the Russian leadership and in the media.

In the second part of the chapter, I trace how the discourse of reconciliation examined in the first part became muted during this period, and suggest that several more 'local' ways of talking about Chechnya, in the Ministry for Internal Affairs and in the FSB (*Federal'naya Sluzhba Bezopasnosti*, the Federal Security Service of the Russian Federation), entered official discourse and contributed to defeat the discourse of reconciliation. I focus on the March 1999 abduction of the Russian President's Envoy to Chechnya, Major-General Gennady Shpigun of the Ministry for Internal Affairs. The official statements accompanying this occurrence contained a securitizing narrative that broke with the discourse of reconciliation. Again, while the main focus is on the linguistic practices, I also comment on the violent practices against Chechnya following the Shpigun case, as well as the authorization of the agencies of violence enabled by this surge of securitizing talk.

Given the meta-theoretical foundation of this study, dramatic real-life events such as abductions, military incursions or terrorist attacks are of particular interest because they present an opportunity to discover how such events are handled and given meaning linguistically. The multitude of statements that are triggered when such events take place can provide a rich reservoir of sources for studying the changing pattern of official representations of Chechnya and Russia. Such real-life events therefore recur throughout the account which follows. This will also offer a rough outline of what happened in the period under study, even if the main focus remains on the shifting pattern of representations.

'Centuries-old confrontation is coming to an end'

Russian President Boris Yeltsin (1991–1999) was not the most prominent voice on Chechnya in the interwar years. He was ill for most of his second term of office, and it is difficult to find statements by Yeltsin on the subject of Chechnya, let alone statements framed in negative terms. However, this silence should be understood as a contribution to the de-securitization process already underway in 1996. On 14 August 1996, Yeltsin signed a decree granting Security Council Secretary Aleksandr Lebed primary responsibility for finding a settlement to the Chechen conflict.[2] This gave Lebed authority to sign the Khasavyurt Accord, which should be recognized as a loud statement in the emerging de-securitizing discourse on Chechnya.[3]

Fairly representative of the official discourse up until the peace deal, which was reached on 31 August 1996, was a representation of Chechnya as an 'Afghan scenario' that necessitated 'ruthless measures against the terrorists and criminals in the Republic of Chechnya'.[4] However, Lebed's statements offered a very different picture, and suggested very different policies. Rebuffing claims that appeasing Chechnya would result in the unravelling of Russia, Lebed argued that Moscow must view Chechnya as 'unique' and not assume that Russian policy toward Chechnya would have automatic resonance elsewhere. He added that Russians should recall their past dealings with Chechens. 'In the last century', he said, 'Russia was unable to defeat the Chechens by force. Diplomacy brought peace. That's how we must act today as well.'[5] Defending the Khasavyurt Accord during the Duma discussion on Chechnya on 3 October 1996, Lebed argued that there was no way to solve the Chechen conflict by force: the war was 'a most stupid war'. Peace was in the interest of Russia. He supported his argument by stating that 'the war had cost between 80,000 and 100,000 lives', thus constructing the Russian-initiated war rather than Chechnya as an existential threat.[6] Strengthening this discourse, Prime Minister Chernomyrdin changed his previous position, now labelling the peace deal a 'success'. Moscow had made mistakes in Chechnya he said, and 'we must speak of our shame for everything that has happened'.[7]

However, there were also those in the Russian leadership who described the peace deal as dangerous. Interior Minister Anatoly Kulikov gave the most vocal contribution to this alternative discourse. During the October 1996 discussions on the Khasavyurt Accord in the Russian Duma and the Federation Council, Kulikov said that Lebed's 'appeasement' of the Chechens rivalled that of British Prime Minister Neville Chamberlain at Munich. He denounced the accords as 'national betrayal' and claimed that the peace deal would only serve the forces that are intent on 'destroying Russia'.[8] In characterizing the Chechen adversary before the Federation Council, he stated that 'terror ... is the basis on which the separatists are building their post-Khasavyurt government'.[9]

Kulikov's securitizing statements were reinforced by those of the political opposition in the Duma. Speakers from three of the four largest parties even accused Lebed of 'high treason'.[10] The CPRF leader Gennady Zyuganov linked Chechnya to the discourse on disintegration that had dominated the Chechnya debate at the beginning of the war in 1994. He warned that Russia was gradually repeating the destiny of the Soviet Union, and that state breakup had begun without anyone noticing. He also appealed to the State Duma to raise awareness of the threat of Chechen separatism as well as the wider atomization of the Russian Far East.[11]

Thus, there were two basic and competing discourses on Chechnya in the Russian leadership at this time. While Lebed was arguing that the Russian Army had to withdraw from Chechnya to save itself, Interior Minister Kulikov held that the army must remain in Chechnya in order to save the state.[12] Yeltsin refrained from explicit pronouncements on Chechnya, although on several occasions he did indicate his support for Lebed's line.[13] Lebed was removed from the post of Security Council Chief on 17 October 1996, but statements by his successor Ivan Rybkin also contributed to strengthening the de-securitizing narrative on Chechnya.[14]

In general, Rybkin's descriptions of and statements on Chechnya helped to lessen the potential for renewed confrontation. Rybkin's language served to humanize the newly elected Chechen president and include him in a Russian 'Self' identified with law and order. Speaking after the inauguration of Aslan Maskhadov in Grozny in February 1997, Rybkin stated that he believed the elected president of Chechnya would be able to protect human rights and the rule of law, thus totally contradicting the image of the Chechen adversary as 'bandits', which was so widespread during the first war.[15] Commenting on the forthcoming May 1997 treaty of peace and friendship between Russia and Chechnya, Rybkin claimed that the 'centuries-old confrontation is coming to an end'.[16] In the treaty, both parties pledged to 'forever refrain from applying or threatening the use of force to resolve any question of controversy'.[17]

However, representations of Chechnya in official statements were very different from those appearing in the Russian press. From the first day of the signing of the peace accord and thereafter, Russian media accounts emphasized hardliner statements by Chechen actors re-asserting Chechen independence (despite agreement not to address this question until 2001). Such accounts reported: attacks by Chechen militants on Russian soldiers; scores of abductions, including the abduction of Russian servicemen, journalists and officials by Chechen field commanders; killings of civilians, including foreign volunteer Red Cross workers and telecom workers; repeated bomb blasts in areas bordering Chechnya; Chechen cross-border raids into Dagestan; the presence of unofficial Sharia guards in Chechnya; the introduction of Islamic dress code; the Chechen government's recognition of the Taliban regime in Afghanistan; the rising power of the Sharia high court and of armed Islamist groups (*jamaats*); the introduction of Sharia law throughout Chechnya; the establishment of a 'Congress of Chechen and Dagestani people' by the radical opposition; and the establishment of Basayev's 'Peacekeeping Brigade' consisting of 'several thousand well-armed fighters' – and the list could be continued.[18]

Taken together, media accounts during the interwar period increasingly represented 'Chechnya' as radically different from 'Russia' – and dangerous for it. This was effectuated not least by the near-total absence of positive images or characterizations of 'Chechnya' and key Chechen actors. My point here is not to suggest that the real-life events reported in the media did not happen – only to note that by this time the image of 'Chechnya' conveyed by the newspapers already coincided with the alternative discourse on Chechnya strongly articulated by Kulikov. On the other hand, a striking feature of the first two years after the signing of the Khasavyurt Accord is the mismatch between these media representations and the absence of securitizing statements by most Russian government representatives.

When confronted with the 'fact' that all presidential candidates in the 1997 Chechen presidential elections supported independence, Prime Minister Viktor Chernomyrdin simply indicated that 'one should not take seriously' campaign rhetoric on independence. 'Let the elections happen and when everything has calmed down we can sit down at the table and begin working together', he said.[19] Yeltsin commented on the particularly brutal killing of the Red Cross workers by saying that he was 'shocked', while Rybkin strongly rejected Chechen allegations that Russian security forces were involved – but no government official blamed the killings on the Chechen leadership.[20] Instead, Rybkin hinted that they might have been committed by people who were opposed to a settlement of the conflict, including 'guests who have been invited and now do not want to leave' – a very 'soft' way of describing the foreign *jihadis* in Chechnya.[21] Concerning many of the violent incidents in Chechnya widely covered in the Russian media in this period, it is difficult to find any comments from the Russian leadership at all.

The exception to this rule was Interior Minister Anatoly Kulikov, who offered statements that represented Chechnya as a threat to Russia for nearly every incident, and sometimes instructed his Ministry to take active measures against the 'Chechen threat'.[22] Kulikov stated that the attack by Chechen militias on a Russian military base in December 1997 justified 'pre-emptive strikes at bandit strongholds wherever they are situated, including on Chechen territory'.[23] He expanded his argument by saying that this was not about starting a new war in Chechnya, but 'a fight against terrorists and bandits ... who have to carry responsibility to the point of their own physical destruction'.[24] Already here, we find what was later to become the official characterization of the Chechen threat and the strategy for dealing with it. At this time, however, Kulikov was reprimanded by Yeltsin, who stressed that the Minister should find better ways of expressing himself so that his statements would not be taken as a call to war.[25] On 25 March 1998, Kulikov was removed from his post by Yeltsin. Rybkin had been suggesting that Russia use only economic pressure on Chechnya, and not launch 'pre-emptive' military strikes. 'Evil leads only to evil', he said, 'especially when a whole nation is punished.'[26]

Even the abduction of the Russian President's envoy to Chechnya Valentin Vlasov on 1 May 1998 did not trigger harsh statements against the Chechen leadership. Rybkin, like Chechen President Aslan Maskhadov and Boris Berezovsky (then CIS executive secretary), condemned the kidnapping as a political act aimed at sabotaging peace talks between Russia and Chechnya and destabilizing the North Caucasus.[27] The new Interior Minister Sergey Stepashin called for a *joint operation* to secure Vlasov's release, and steps were taken to increase cooperation between the Russian and Chechen Interior Ministries. A Russian Interior Ministry mission was re-opened in the Chechen capital, and 60 Chechen police officers were invited to attend a training course in Moscow.[28]

Instead of seizing the opportunity to blame the precarious security situation and deteriorating negotiating atmosphere on the Chechen leadership, the Russian leadership seemed to be arguing that Russia was to blame, and that the solution was to be found in continued economic support to the Maskhadov regime. For example, echoing a statement made by Yeltsin earlier, Rybkin conceded that 'Russia is not doing very well' in honouring earlier agreements signed with Grozny, including pledges to provide economic assistance.[29] Such expressions of faith in economic support as the solution to the Chechnya problem continued even during autumn 1998, when civil war seemed to be threatening Chechnya following the July assassination attempt against Maskhadov and ensuing violent clashes between forces loyal to Maskhadov and the armed radical opposition. This viewpoint – that honouring earlier economic agreements and creating a free economic zone in Chechnya would save the Maskhadov regime – was expressed by Prime Minister Kiriyenko, Boris Berezovsky, former Prime Minister Viktor Chernomyrdin, and the president of Tatarstan Mintimer Shaymiyev.[30]

In August 1998, Kiriyenko again indicated that Russia had failed to implement bilateral agreements, adding: 'we need peace and stability in the North Caucasus ... we need to find solutions to the economic problems of the Chechen Republic of Ichkeriya and the neighbouring regions'.[31] Yevgeny Primakov signalled that he would devote more attention to the problems in North Caucasus when he became Prime Minister in September 1998, but what he proposed was to continue to channel more money from the centre in order to alleviate the socio-economic problems in the region, and to support and cooperate with the Maskhadov regime.[32] During his meeting with Maskhadov in October 1998 Primakov promised that 'Russian Federation military forces would never again be brought into Chechnya'.[33]

Within this enduring de-securitizing discourse, the Russian leadership presented a consistently positive image of President Maskhadov. Typical here is Rybkin's description of Maskhadov as 'the legitimately elected president and the guarantor of stability in the region'.[34] Similar pronouncements came also from President Yeltsin, Prime Minister Primakov, Interior Minister Stepashin and even the heads of law enforcement structures.[35]

However, during autumn 1998 a clear distinction emerged in the Russian official discourse on Chechnya between Chechnya as the Maskhadov regime on the one hand, and those who were increasingly represented as the culprits behind the violence and chaos, referred to as 'terrorists', 'bandits', 'extremists', 'hotheads' or 'foreigners' on the other. Interestingly, this rhetoric emerged in the period following the August 1998 bombing of U.S. embassies in Kenya and Tanzania, when Saudi-born Osama bin Laden was singled out as the man behind the terrorist attacks. True, many of these characterizations have a long history in the Russian discourse on Chechnya, but the re-emergence of some of these labels at this particular time suggests an emerging intertextual relationship between the U.S. discourse on international terrorism and the Russian discourse on Chechnya.

Not only did Russian official discourse now hint at a religious dimension to the internal Chechen conflict, it also indicated that the threat came from abroad. Rybkin suggested in September 1998 that former Chechen president Zelimkhan Yandarbiyev, together with 'extremist Islamic circles' in the Middle East, were backing the Chechen field commanders who were threatening Maskhadov's position. He said that 'certain circles in the Middle East who propagate extreme forms of Islam' had taken advantage of the fact that Zelimkhan Yandarbiyev had never come to terms with his failure to be elected as president in 1997.[36] Also, Rybkin later blamed violence in Chechnya on 'outsiders from Jordan and Saudi Arabia'.[37] In the Duma, the chair of the Committee on Nationality Issues Vladimir Zorin defined the divide within Chechnya as a religious one, and said that Maskhadov was facing opposition because he had raised the 'banner of anti-Wahhabism'.[38]

Russian press coverage at this time contributed to locating Chechnya within the realm of a rising anti-terrorist agenda in the international community. Although I have found no references by Russian leaders directly linking the radical opposition in Chechnya to Osama bin Laden at this time, as early as 22 August 1998 *NeGa* had indicated that Osama bin Laden or his representative had visited Chechen soil – adding that it was unnecessary to explain 'what results such visits could bring'.[39] Referring to the Arab newspaper *Hayat*, in December 1998 *NeGa* suggested that Osama bin Laden was to be accorded refugee status in Chechnya.[40]

As 1998 was drawing to a close, official Russian statements abandoned the de-securitizing discourse and turned silent on Chechnya, accompanied by an increasing lack of action. A meeting planned for November 1998 between the Russian Minister of Interior and Chechen leaders in North Caucasus never took place. In December 1998, Yeltsin annulled the September 1997 directive that allowed the drafting of a treaty with Chechnya on the mutual delegation of powers with the Russian Federation. Yet another commission was established to address the Chechen problem, but there was no accompanying information campaign. The initiative was followed up by pledges of increased economic assistance to Chechnya and strengthened mutual cooperation between Russian and Chechen law enforcement agencies in the 'fight against crime', but the results were meagre.[41]

In the meantime, the Russian press increasingly depicted Chechnya as a scene of internal chaos and danger, with the potential to impact negatively on Russia, and criticized the Russian authorities for not taking action to defend Russia's interests.[42] These articles clearly portrayed the conflict in Chechnya as a religious conflict and did not hesitate to place the terrorist label on the Chechen leadership. [43]

'Russia has run out of patience with the process of ever-deepening criminalization of the republic'

The abduction of the Russian Presidential Envoy to Chechnya Major-General Gennady Shpigun in Chechnya on 5 March 1999 was followed by a change in the official discourse on Chechnya. Most official statements as well as statements from across the political spectrum can be read as part of a securitizing move. Together they constructed 'Chechnya' as highly dangerous, and indicated that violent retribution was required. And here we should recall that the discourse accompanying the comparable abduction of Valentin Vlasov in May 1998 had been a totally different one.

In a statement issued on 7 March 1999, Minister for Internal Affairs Sergey Stepashin said that, despite assurances from the Chechen leadership that it was cracking down on crime and terrorist activities, 'Russia has run out of patience with the process of ever-deepening criminalization of the republic.' He warned that Moscow would resort to 'extremely rigorous measures to ensure law, order, and security in the North Caucasus region', including the annihilation of the bases of 'bandit formations' on Chechen territory.[44] 'In effect, several thousand armed scoundrels dictate their will to Chechen society, driving it [the territory] into medievalism and obscurantism', Stepashin declared.[45]

Stepashin's characterizations drew on the discourse that had been prevalent in the Ministry for Internal Affairs during the interwar years but had lost out at the official level.[46] It is also noteworthy that, during Stepashin's time as Minister for Internal Affairs, several initiatives in Chechnya had already been launched under the label of 'anti-terrorism'.[47]

Stepashin's statements were followed up by First Deputy Prime Minister, Vadim Gustov, who declared that a 'direct challenge has been thrown at Russian power' and promised that a solution would be found – 'even if it was one that would not be very popular'.[48] Security Council Deputy Secretary Vyacheslav Mikhaylov also supported the 'destruction of bandit formations on Chechen territory'. Drawing a parallel to the U.S. bombings in Afghanistan, he noted that there existed 'a worldwide practice' (*mirovaya praktika*) on how to deal with 'terrorist training camps'.[49] Head of the Duma Security Committee Viktor Ilyukhin said 'it is well known where in Chechnya the terrorist bases are located', and proposed eradicating them through the use of air power.[50]

There were, however, also several official statements more in line with the 'discourse of reconciliation'. Prime Minister Yevgeny Primakov concluded that 'force would not be used to solve the Chechen conflict', and that the President thought it 'of utmost importance to keep peace in Chechnya'.[51] Also, Stepashin's comments on Chechnya decreased markedly after he became Prime Minister on

13 May 1999. Nevertheless, the few statements he gave on Chechnya were fairly consistent with those accompanying the Shpigun abduction. In connection with the freeing of two Orthodox priests who had been kidnapped in Chechnya, Stepashin personally met them and stated: 'Such scoundrels who abduct people should not only be punished, they should be annihilated; there is no place for them on this earth'.[52] Later on, responding to the Ministry for Internal Affairs's reports on the situation on the Chechen border, Stepashin issued instructions to 'take exhaustive measures to annihilate bandits who endanger the life and health of Russian citizens and representatives of the organs of power and government'.[53]

Thus, a struggle between competing positions on Chechnya reached the official level in this period. On balance, the discourse accompanying the Shpigun case represented a rupture with the dominant official discourse of the interwar years, and was related to the discourse that was subdued but kept alive by Kulikov and the media. Even if President Maskhadov was not included in these representations, they linked 'Chechnya' to 'bandit', 'crime' and 'terror' and proposed 'annihilation' and 'destruction' as relevant policies for dealing with Chechnya. Moreover, the emerging securitizing discourse explicitly drew on a narrative which existed in the international discursive context – the reference to 'worldwide practice' on how to deal with 'terrorist camps'.

Policies on Chechnya also changed. After the Shpigun abduction, the Dagestani border with Chechnya was effectively closed. All federal Russian representatives had left Chechnya by 7 March 1999 and no one was appointed to take over the post as Presidential Envoy to Chechnya.[54] Moreover, the change in official statements on Chechnya following the Shpigun abduction made it possible and legitimate to shift dealings with the republic over to the 'power-ministries', and to deal with Chechnya under a rubric other than 'cooperation' or 'economic assistance'. As Head of the Russian Federation Security Council, Vladimir Putin (appointed on 29 March 1999) was made responsible for continuing the work on 'preparing and accomplishing negotiations on the regulation of relations between Russia and Chechnya'.[55] However, instead of continuing the work of negotiating, the Russian President signed a decree prepared by Putin 'On additional measures in the fight against terrorism in the North Caucasus' on 19 May 1999, signalling that Chechnya was no longer to be dealt with under the heading of negotiation, but under the heading of anti-terrorism.[56] The new Minister for Internal Affairs Vladimir Rushaylo was reported as immediately making implementation of this decree the top priority of his ministry.[57]

Shortly after Sergey Stepashin was appointed Prime Minister in May 1999, Russian forces attacked positions inside Chechnya for the first time since 1996. On 28 May 1999, Russian helicopter gunships belonging to the Ministry for Internal Affairs launched airstrikes against a small island in the Terek River that spokesmen claimed was the site of a 'Chechen terrorist camp'.[58] Similarly, the Russian military responded to rebel attacks on Dagestani militiamen close to the Chechen border on 3 June 1999 with air strikes on targets within Chechnya itself. According to official figures on rebel casualties during this period, around '200 Wahhabis' were killed by Russian forces.[59] At the same time, 70,000 Russian troops (counting troops of the Ministry for Internal Affairs, the Ministry for Defence and border troops) were already concentrated along the Chechen border by July 1999.[60]

Russian forces had also started to use mortars and artillery in their attacks on Chechen fighters. On 5 May 1999, after a meeting with the Russian President, the Minister for Internal Affairs Vladimir Rushaylo announced that Russian forces had conducted 'pre-emptive strikes' against fighters on the Chechen–Dagestani border.[61] We may recall that threats of similar retaliation made by Minister for Internal Affairs Kulikov in December 1997 had seemed alien in the dominant discourse at the time: Kulikov was reprimanded by the president and later dismissed.

The claim made here is not that the new representation of Chechnya can *explain* the new Russian approach to Chechnya. Individual violent responses might well have been launched without these official securitizing statements – indeed, that had sometimes been the case during the interwar period. At the time, such instances of violent response had seemed appropriate to the actors who undertook them because they were in line with the 'local' discourse on Chechnya in the Ministry for Internal Affairs, but they did not make much sense within the dominant official 'discourse of reconciliation'. The significance of the new accumulation of official statements with a securitizing narrative lies in the fact that it served to constitute policies of violence as both appropriate and legitimate.

Another aspect of this process is that the strengthening in official statements of a securitizing narrative seemed to authorize certain political actors and establishments that had had an uneasy fit with the 'discourse of reconciliation' – such as those from the 'power-ministries' and the FSB. Many noted that Stepashin's tough rhetoric on Chechnya in connection with the Shpigun abduction and his proposal of harsh policies for dealing with Chechnya made him stand out as the most vigorous and active figure in the Russian leadership, someone really willing to tackle problems. Thus, changes in the official representation of Chechnya served to increase Stepashin's authority, temporarily at least.

Moreover, the emerging discourse focusing on Chechnya as a security threat (re-)constituted the 'power-ministries' and the FSB as relevant and authoritative agencies in the Russian political system. The changing pattern of representations boosted the influence of these agencies and their representatives within the Yeltsin regime in general, and in particular when it came to the handling of the Chechen issue. The very *raison d'être* of these agencies is to deal with security problems, and the urgent security situation brought into being by the emergence of a securitizing narrative in official statements made them key actors.

Experts agree that Yeltsin relied increasingly on the *siloviki* (Russian word for politicians from the military and security services) in this period of deep crisis of confidence in the Russian leadership.[62] The 1998 financial crisis, followed by grave social and economic problems for Russian citizens, and accompanied by the frequent changing of prime ministers[63] and the public absence of Yeltsin for long periods due to bad health, created the general impression of a leadership not capable of solving key issues. Indeed, the opposition had no difficulty in focusing *their* securitizing efforts on the danger posed to the country by the Yeltsin regime. CPRF leader Gennady Zyuganov predicted a 'social explosion' if the government did not take steps to secure a better life for ordinary Russians.[64] In addition, Yeltsin had an impeachment case to deal with in May 1999. In this situation, the new-style official discourse on Chechnya in connection with the Shpigun abduction seemed to enhance the authority of the Yeltsin regime, giving it a much-needed touch of decisiveness and strength.

However, the securitizing move which started with official statements on the Shpigun case was quite 'weak' in terms of the number of statements. While Stepashin's statements initially linked Chechnya to terms such as 'terrorism' and 'bandit' and favoured policies such as 'annihilation' and 'destruction', his later statements were more in line with the 'discourse of reconciliation'. On 23 July 1999, Stepashin declared that there would be no further war in Chechnya, stating that 'Nobody wants to repeat the same mistake twice', and reaffirming that a meeting between the Chechen and Russian presidents would take place. He also dismissed suggestions that the Chechens were a uniquely criminal group: 'Bandits have no nationality and we must fight against them as against bandits and not as representatives of this or that nationality'.[65]

When fighting erupted in the Tsumadin and Botlikh districts of Dagestan after an attack from across the Chechen border by so-called 'Wahhabis' from Dagestan, Chechnya and also other Muslim countries in the first days of August 1999 (events later referred to as the 'invasion of Dagestan'), comments from the Russian Premier Stepashin were few and far between. Although Stepashin noted that 'we may lose Dagestan' in response to the fact that the 'Wahhabis' were in control of six Dagestani villages, he had only days before ruled out a new war in Chechnya. The absence of forces from the Ministry for Internal Affairs or Russian Army troops to fight off the attack in this part of Dagestan was conspicuous.[66]

The lack of official verbal response to the 'invasion of Dagestan' and the failure of the Russian Army and the Ministry for Internal Affairs's troops in Dagestan in early August was sharply criticized in the Russian press. Noting the killing of policemen and civilians before Russian forces finally arrived, the press drew parallels to Russian failures in the First Chechen War.[67] Moreover, press accounts accentuated the radical Islamic aspect of the Chechen threat, repeatedly using words like 'Islamist', 'Sharia', 'terrorist' and 'Wahhabi', and presented what Russia was facing in the North Caucasus as a 'full-scale war' against such forces.[68] Thus, as noted earlier, representations of Chechnya in the Russian press contributed to constructing Chechnya as a radical Islamic threat, even if this version did not dominate the official discourse at the time.

Within the FSB, a discourse on Chechnya similar to that in the Russian press was probably dominant during the interwar period, although the sources here are few. To take one example, a document on Chechnya leaked from the FSB to the press in the interwar period construed the separatists as co-opted by 'foreign Islamic forces': not only were the separatists presented as funded primarily by 'international Islamic organizations', 'foreign special services' and 'Muslim states', but their ranks were represented as filled with 'foreign Islamic fighters' from 'Jordan, Lebanon, Turkey, Azerbaijan, Saudi Arabia and Afghanistan'. According to this FSB document, fighters from 'Afghanistan and Pakistan' functioned as 'instructors' and 'commanders' among the Chechen fighters who were also trained in 'Afghanistan, in Khost and in Peshavar, Pakistan', and young Chechens aged 14–16 had been sent for religious education in 'Jordan, Turkey, Saudi Arabia and Syria'. The words 'Islam', 'Wahhabi' and 'terror' recurred throughout the document, which ended by linking this foreign involvement to the 'eternal' 'criminal' and 'corrupt' 'Chechen mafia'[69] – the latter terms being part of the dominant discourse on Chechens in the 1990s.[70]

Examples of this linking of 'Chechnya' to 'terror' and 'crime' can also be found in the language of Vladimir Putin as FSB Chief (July 1998–March 1999).[71] In the days following the incursion into Dagestan, the FSB informed the Russian press that they had information that the saboteurs from Chechnya were planning terrorist acts in Moscow, St. Petersburg, Makhachkala and Vladikavkaz. This seems to indicate that, although the official discourse in the interwar period was one of 'reconciliation', a quite different discourse on Chechnya as 'crime' and 'terrorism' prevailed in the FSB. With the change of prime ministers on 9 August 1999, representations resonating with this discourse moved to the top official level.

Conclusion

Official Russian discourse on Chechnya moved from de-securitization through silence to budding re-securitization during the course of the interwar years, 1996 to 1999. In the first period, dominated by what might be termed a 'discourse of reconciliation', Chechnya was represented as a victim rather than a threat or a culprit. 'Maskhadov' was linked to terms such as 'human rights', 'rule of law', 'legitimate' 'and guarantor of stability'. In this de-securitizing narrative, there was hardly any degree of threat implied in the concept of Chechnya. Moreover, the narrative indicated that continued use of force would result in the most destructive future for Russia. In line with this argumentation, the policies put forward were 'diplomacy', 'refraining from the use of force', 'cooperation' and 'economic assistance'. Chechnya was indeed 'shifted out of emergency mode and into the normal bargaining processes of the political sphere' in the terminology of securitization theory.

'Russia', within this discourse of reconciliation, was linked to fairly new (in the Russian context) terms such as 'shame', 'blame' and 'failure', all contributing to the construction of the Russian Self as a humble, repentant entity. Descriptors such as 'unable to defeat' and 'failure to implement' also constructed Russia as a weak actor. In short, if there were anyone to blame for the problems in Chechnya, it was Moscow.

This de-securitizing narrative did not determine the policies pursued by Russia toward Chechnya in this period, but it did open up the possibility of pursuing policies such as the total withdrawal of Russian troops from Chechnya, the signing of a peace treaty in May 1997, economic assistance,[72] attempted security cooperation through joint police training and joint commissions on crimes and on money-laundering. The boundaries that had been drawn between 'Chechnya' and 'Russia' in the de-securitizing narrative now made such policies appear both logical and legitimate.

Yet, the set of texts reviewed here also indicated that the discourse of reconciliation never acquired a hegemonic position, even in the texts of the political leadership. Struggling with and temporarily outdone by this discourse of reconciliation was another basic discourse on Chechnya, strongly articulated by Interior Minister Kulikov and by the political opposition as well. This discourse linked Chechnya to terms such as 'criminal', 'bandit' or 'terrorist' and constructed any peace deal with Chechnya as an existential threat to Russian statehood. Policies of negotiation and compromise were represented as 'national betrayal' and 'appeasement', thus linking them to Russia's historical experiences with the Nazi threat. In this narrative, policy suggestions included 'pre-emptive strikes' and 'physical destruction'. In line with this narrative, policies such as the deployment of troops seemed legitimate and were actually pursued by the Ministry for Internal Affairs on its own initiative in this period.

Even if this securitizing discourse was subdued by the dominant discourse on Chechnya as a victim in official rhetoric, it was kept alive by representations in the media. The accounts of the Russian media clearly framed developments in Chechnya as threatening to Russia. By the end of the interwar period, 'Chechnya' had become linked to 'Bin Laden' and the international terrorist threat. The conflict was represented as one of civilizations. Against this background we could say that the 'securitizing move' actually started in Russian media accounts rather than in the statements of the Russian leadership – which testifies to the intersubjective nature of securitization. As noted in the theory chapter, the securitizing process cannot be understood as top-down and unidirectional. It has to be understood as a joint act. Also, our examination of the Shpigun case showed that official securitizing moves can be fed by discourses that prevail in 'local' official constituencies. Not only media accounts, but also those of the Ministry for Internal Affairs and the FSB long before summer 1999, offered representations of Chechnya and Russia that resembled those that later emerged as the core of a new official securitizing narrative.

During the autumn of 1998 there was a shift in official statements, towards singling out 'the Chechen extremists' as a threat to the Maskhadov regime and to the peace process. This threat was given as directly linked to, even fostered by, Islamic extremist circles in the Middle East. Picking up on the media discourse, the internal divide in Chechnya was represented as concerning issues of religion. This new trend in Russian official language on Chechnya also seemed to be fostered through yet another inter-textual relationship. The emerging U.S. securitization of the terrorist threat following the bombing of the embassies in Kenya and Tanzania constituted this discourse as significant. With time, the U.S. official discourse would privilege and advantage a Russian official discourse on Chechnya that turned the interwar 'discourse of reconciliation' on its head – but that is a different story.

Findings on the Shpigun case also foreground the topic that this book will address in depth in Chapters 10, 11 and 12, namely that securitizations are not limited to linguistic machinations. They enable and constrain policies and material practices. With the new way of talking about Chechnya introduced in statements on the Shpigun case, violent actions such as the bombing of Chechen territory and the concentration of troops around the Chechen border were again rendered logical and appropriate.

Finally, this chapter also illustrates a social mechanism presented in the theory chapter: how securitizations authorize actors to address the existential threats that they bring into being. The agencies working in the field of security and those that administer violence were immediately given a prominent position and brought back to the centre of Russian politics as early as spring 1999 and as the securitizing statements on Chechnya re-emerged.

Notes

1 M. Lermontov, 1977. Cossack lullaby. In: *Tragedy in the Caucasus*, translated by L. Kelly. London: Constable, 207.
2 The decree gave Lebed additional powers to coordinate the activities of federal executive agencies and dissolved Prime Minister Viktor Chernomyrdin's State Commission for Regulating the Chechen Conflict (*RFE/RL Newsline*, 15 August 1996).
3 Chechen Chief of Staff Aslan Maskhadov and General Aleksandr Lebed signed an agreement on 'Principles for the determination of the basis of relations between the Russian Federation and the Chechen Republic' in Khasavyurt on 31 August 1996. The text of the agreement 'Khasavyurtovskiye soglasheniya' was published in *NeGa* on 3 September 1996.
4 Russian Prime Minister Viktor Chernomyrdin, quoted in 'Chechnya: Protivoborstvuyushchiye storony zagnali drug druga', *NeGa*, 9 August 1996.
5 Cited in 'Is Chechnya Tet or Tatarstan?' *RFE/RL Features*, 14 August 1996.
6 'Glavnoy temoy pervogo zasedaniya Dumy stala natsional'naya bezopasnost', *NeGa*, 3 October 1996 and *RFE/RL Newsline*, 4 October 1996.
7 'Chernomyrdin odobril deystviya Lebedya', *NeGa*, 3 September 1996 and *RFE/RL Newsline*, 4 September 1996.
8 'Glavnoy temoy pervogo zasedaniya Dumy stala natsional'naya bezopasnost', *NeGa*, 3 October 1996 and 'Russia: Analysis from Washington: The Chechen war resumes in Moscow', *RFE/RL*, 4 October 1996.
9 Speech of the Minister for Internal Affairs of the Russian Federation Kulikov in the Federal Council, 8 October 1996, referred to in *NeGa*, 11 October 1996. Kulikov had earlier warned that, if Russian troops were withdrawn completely, a campaign of terror would be unleashed against all Chechens who cooperated with Moscow: see *RFE/RL Newsline*, 12 September 1996.
10 'Russia: Analysis from Washington – the Chechen war resumes in Moscow', *RFE/RL*, 4 October 1996.
11 'Glavnoy temoy pervogo zasedaniya Dumy stala natsional'naya bezopasnost'', *NeGa*, 3 October 1996 and in *RFE/RL Newsline*, 3 October 1996.
12 Jacop Kipp 'Experts see dangers in military problems', *RFE/FL Features*, 13 December 1996.

13 For example, in his remarks at the end of the year in 1996, Yeltsin said that the 'past year was marked by the establishment of peace in Chechnya' and that 'this was the President's line' (President ob ukhodyashchem i novom gode', *NeGa*, 31 December 1996).

14 In May 1997, Yeltsin ordered all government and state structures to coordinate their statements and actions on Chechnya with Rybkin. Thus, Rybkin was, so to speak, given the right to form the official Russian image of Chechnya. 'Polnomochiya Ivana Rybkina sushchestvenno rasshireny', *NeGa*, 6 May 1997.

15 'Rybkin defends legitimacy of Chechnya's President', *RFE/RL Features*, 12 February 1997.

16 'Polnomochiya Ivana Rybkina sushchestvenno rasshireny', *NeGa*, 6 May 1997.

17 'Boris Yeltsin i Aslan Maskhadov skoropostizhno podpisali dogovor', *NeGa*, 13 May 1997.

18 See, for example, news reports on Chechnya in *Itar Tass*, 27 October 1996; *RFE/RL Newsline*, 31 October 1996, 11 and 18 November 1996, as well as 5, 19 and 20 December 1996, 29 April 1997, 27 May 1997, 11 November 1997, 9, 22 and 30 December 1997 and 24 August 1998; and from *NeGa*: 'Terror vryad li prekratitsya', 1 December 1998; 'Shariatskiy sud nakonets vyskazalsya', 6 December 1998; 'Dagestanu grozit opasnost'', 13 January 1999; 'Voyna v Chechne poka otmenyayetsya', 23 January 1999; and 'Vstrecha Yeltsina s Maskhadovym gotovitsya', 21 April 1999.

19 *RFE/RL Newsline*, 28 January 1997.

20 *RFE/RL Newsline*, 18 January 1996.

21 Rybkin cited in 'Chechens to guarantee safety of international monitors', *RFE/RL Features*, 30 December 1996.

22 For example, the Interior Ministry imposed stricter controls along the border between Russia and Chechnya, to 'prevent thousands of Chechen gunmen from crossing it' (*RFE/RL Newsline*, 19 June 1997). Another example is the deployment of Interior Ministry troops to protect the border in November 1997 (*RFE/RL Newsline*, 14 November 1997).

23 'Kulikov's Chechnya rhetoric stirs fury', *The Moscow Times*, 9 January 1998.

24 *Nevskoye Vremya*, 20 January 1998, article 5.

25 'Russia-President-Chechnya', *Itar-Tass*, 9 January 1998; *Nevskoye Vremya*, 20 January 1998, article 5.

26 *RFE/RL Newsline*, 2 February 1998.

27 *RFE/RL Newsline*, 4 May 1998.

28 *RFE/RL Newsline*, 15 May and 1 June 1998.

29 *RFE/RL Newsline*, 4 December 1997.

30 'Na vstreche v Nazrani Maskhadov budet trebovat' ekonomicheskoy pomoshchi Rossii', *NeGa*, 31 July 1998; and 'Svobodnaya zona i zashchita ot neye', *NeGa*, 4 August 1998.

31 *RFE/RL Newsline*, 3 August 1998.

32 'Yevgeny Primakov nanës vizit v Ministerstvo inostrannykh del', *NeGa*, 26 September 1998; and 'Primakov kontsentriruyet vnimaniye na yuge', *NeGa*, 3 October 1998.

33 'Primakov i Maskhadov na 'perelomnom momente', *NeGa*, 31 October 1998.

34 'Stabli'nost' v Chechne pod ugrozoi', *NeGa*, 29 September 1998.

35 'V neytralizatsii Maskhadova zainteresovany v Chechne mnogiye', *NeGa*, 24 July 1998; *RFE/RL Newsline*, 10 November 1998; and 'Stabil'nost' v Chechne pod ugrozoy', *NeGa*, 29 September 1998.

36 'Stabil'nost' v Chechne pod ugrozoy', *NeGa*, 29 September 1998.

37 *RFE/RL Newsline*, 27 October 1998.

38 'Vnutrichechenskiy konflikt usugublyayetsya', *NeGa*, 10 October 1998.

39 'Chechnya priznala Talibov', *NeGa*, 22 August 1998.

40 'Ben Ladena gotovy prinyat' v Chechne?', *NeGa*, 8 December 1998.

41 'Grozny nachal poluchat' pomoshch'', *NeGa*, 8 December 1998.

42 'V Stavropole gotovyatsya k otrazheniyu diversii', *NeGa*, 17 February 1999.
43 See for example 'Chechnya priblizilas' k khaosu', *NeGa*, 27 January 1999; and 'Budet li Chechnya upravlyat' Rossiyey', *NeGa*, 6 April 1999.
44 'Moskva pytayetsya nayti adekvatnyy otvet deystviyam Chechenskikh ekstremistov', *NeGa*, 10 March 1999.
45 *RFE/RL Newsline*, 9 March 1999.
46 Stepashin headed the FSB from 1994 to 1995, then served in the Presidential Administration and later as Justice Minister from 1997, before becoming Minister for Internal Affairs in 1998.
47 'Stepashin otpravilsya na Kavkaz', *Ria Novosti*, 28 April 1999; 'Real'nost' silovykh mer', *NeGa*, 8 April 1999.
48 'Moskva pytayetsya nayti adekvatnyy otvet deystviyam Chechenskikh ekstremistov', *NeGa*, 10 March 1999.
49 'Primakov prizvan reshit' nerazreshimuyu problemu', *NeGa*, 11 March 1999.
50 'Moskva pytayetsya nayti adekvatnyy otvet deystviyam Chechenskikh ekstremistov', *NeGa*, 10 March 1999.
51 'Prezident za mir s Chechney', *NeGa*, 12 March 1999.
52 'Osvobozhdeny ocherednyye zalozhniki', *NeGa*, 29 May 1999.
53 'Na granitsakh Chechni razvernulas' nastoyashchaya voyna', *NeGa*, 19 June 1999.
54 'Vlasov stal chlenom TsIK', *NeGa*, 19 March 1999.
55 'Yeltsin vstretilsya s Putinym', *Ria Novosti*, 1 June 1999.
56 'Severokavkazskiy krizis proverit pravitel'stvo na prochnost'', *NeGa*, 20 May 1999.
57 'Rushaylo provel soveshchaniye po Severnomu Kavkazu', *NeGa*, 25 May 1999.
58 *RFE/RL Newsline*, 31 May 1999.
59 'Na granitsakh Chechni razvernulas' nastoyashchaya voyna', *NeGa*, 19 June 1999.
60 'Po Chechne nanosyatsya preventivnyye udary', *NeGa*, 6 July 1999.
61 Ibid.
62 Public opinion polls clearly showed that Russian citizens had no confidence in their president: one poll from May 1999 showed that only 2.2 per cent of the respondents had full confidence in the president ('Strana ne doveryayet svoyemy prezidentu', *NeGa*, 7 May 1999).
63 Chernomyrdin was replaced by Kiriyenko on 24 March 1998; on 23 August, Chernomyrdin was selected to take over again but was rejected by the Duma, so on 11 September Primakov was confirmed as Prime Minister – only to be replaced by Stepashin on 13 May 1999.
64 'Zyuganov predrekayet sotsial'nyy vzryv', *NeGa*, 17 February 1999; 'Posledniy Oligarkh ukhodit?', *NeGa*, 6 March 1999.
65 *RFE/RL Newsline*, 23 July 1999.
66 'Rossiya prodolzhayet nastypat' na grabli', *NeGa*, 10 August 1999.
67 Ibid. Also 'Islamskaya revolyutsiya v Dagestane', *Kommersant*, 4 August 2008.
68 'Soprotivleniye Allakhu', *Vremya MN*, 24 August 1999. 'Imamat xxi veka', *MK v Pitere*, 19 August 1999; 'Full scale war', from 'Segodnya Dagestan otdelyat ot Rossii', *Kommersant*, 10 August 1999.
69 The document was not published in full, but extensive excerpts were referred to in 'Den'gi dlya Ichkerii', *Interfax AiF*, 9 December 1996.
70 J. Russell. 2005. Terrorists, bandits, spooks and thieves: Russian demonisation of the Chechens prior to and since 9/11. *Third World Quarterly*, 26(1): 101–116.
71 For example, when a bomb exploded in Vladikavkaz on 19 March 1999, killing over 60 people, FSB Chief Vladimir Putin informed the President that, although the FSB was operating with several alternative hypotheses, there 'clearly was a "Chechen trace"' ('Moskva ne zabyvayet o Kavkaze', *NeGa*, 26 March 1999). Later he said that the identity of the perpetrators had been established, but he did not disclose their nationality. In September 1999, however, a Georgian was detained on suspicion of having participated in the bombing, and finally, in 2003 two Ingush

from the North-Ossetian-disputed Prigorodny region went on trial for the bombing (*RFE/RL Newsline*, 7 and 27 April 1999, and then 10 April 2003).

72 The Russian government allocated 847 billion rubles to Chechnya between January and August 1997 (*Interfax*, 17 August 1997.) On 5 August 1997, Russian Security Council secretary Ivan Rybkin said another 700,000 million roubles ($120.7 million) could be delivered to Chechnya before the end of the year (*RFE/RL Newsline*, 5 August 1997).

5 Russian official representations of Chechnya and Russia

On 9 August 1999, President Yeltsin decided to sack Stepashin and appoint Vladimir Putin as Prime Minister. Putin later recalled that, in relation to his appointment and the incursion into Dagestan, he saw it as his mission 'to bang the hell out of those bandits'.[1] When meeting with Russian journalists the day before the Duma was to vote over Putin's candidacy, Yeltsin confirmed his support for Putin and promised that emergency rule would not be introduced, but added that 'the toughest measures possible were necessary to install order in the North Caucasus'. He believed that Putin was the right man to achieve this goal.[2] In his speech to the Duma the following day, Putin underlined his intention to introduce 'the toughest measures possible' to deal with the violent conflict in Dagestan.[3]

Among Putin's first moves as Prime Minister was to arrange a meeting on Dagestan in the Russian Security Council. He also summoned a meeting of the Federal Antiterrorist Commission and opened the meeting in the Commission, where all heads of Russia's power ministries and departments were present, by declaring, 'In the Caucasus and in Dagestan specifically we are facing lawlessness and terrorism. This is a situation we cannot tolerate on Russian territory.'[4] Further: 'Yesterday I ordered the Ministry for Internal Affairs to establish order and discipline there.'[5] A plan for a military operation in Dagestan, subsequently approved by Yeltsin, was worked out, and Putin announced that 'the situation in Dagestan would be straightened out in one-and-a-half to two weeks'.[6] After a meeting with Putin, Minister for Internal Affairs Vladimir Rushaylo confirmed, 'we will make it within the deadlines set for the counter-terrorist operation'.[7]

Thus, in the course of only a few days, the threat facing Russia in the North Caucasus became the top issue in Russian politics. It was clearly represented as 'lawlessness' and 'terrorism', in turn making the 'toughest possible measures' the most logical response. And Putin was projected as the man capable of launching such a response.

Much has been said and written about Vladimir Putin's masterful propagation of the Second Chechen War. Indeed, observing this process and the transformation of Russia that ensued through media reports and personal encounters was what got me started on this project. It was remarkable how

Vladimir Putin's powerful talk, presenting Russia as faced with a terrible Chechen terrorist threat, became a driving force in Russian politics. In securitization theory, such powerful security talk is identified as the starting point of enquiry and the centre of analysis.

This chapter maps out the emergence and strengthening of an official position that served to discursively reconstruct and normalize 'Chechnya' as a 'terrorist threat', by analysing statement after statement during summer and autumn 1999. In the first part, I review the official statements immediately following the appointment of Vladimir Putin as Prime Minister, looking at the key statements on the incursion into Dagestan in August 1999 and the terrorist bombings in Russian cities in late August and early September 1999 – in chronological order, so that it is possible to follow the official contribution to the emerging securitizing move. Some analysis of the statements is presented, but the deeper analysis of the linguistic aspect of the discourse of war follows at the end of this chapter. In the second part, I examine and then organize these official statements according to the analytical template, 'the securitizing narrative', introduced in the theory chapter.

Going through the details of the official securitizing narrative will make it possible to determine the level of threat linked to 'Chechnya', how Russia was represented in the face of this threat, and what policies were suggested to be undertaken against Chechnya. The linguistic practices presented in this chapter acquired a logical material expression in the emergency practices undertaken against Chechnya that are represented in Chapters 10, 11 and 12 and will be revisited in those chapters to demonstrate the co-constitutive nature of linguistic and material practices in a process of securitization. In concluding this chapter, I compare the 1999 narrative to the official narrative that accompanied the launching of the First Chechen War in 1994 and discuss the wider implications of official representations of Chechnya, not only for the republic but also for the Chechens as a people.

'We want to end, once and for all, the centre of international terrorism in Chechnya'

In Russian official statements on the war in Dagestan, certain descriptors were used again and again – 'bandit', but even more frequently 'terrorist' and 'Wahhabi'. For example, when Russian Federal forces moved in to take control over the villages of Karamakhi and Chabanmakhi on 29 August 1999, these villages were referred to as 'Wahhabi' or simply 'terrorist villages'.[8] Note that Stepashin had declared, after visiting Karamakhi a year earlier, that 'the inhabitants here are not at all Wahhabis'.[9]

Another key feature of official language during the brief war in Dagestan in 1999 was the argument that the threat emanated from 'Chechen' territory. At the beginning of the 'large-scale operation to expel terrorists from Dagestan', Putin promised that 'bandits would be targeted anywhere, wherever they might be', and he specifically mentioned 'Chechen' territory.[10] Putin also described the attackers with whom the 'Dagestani nation' had had to fight as simply 'Chechen fighters', although very many of them were from Dagestan.[11] Subsequently, Igor Zubov, the representative of the Interior Minister, confirmed that 'massive anti-terror operations' would be conducted across Dagestan, and that they would pursue the fight into 'Chechen' territory.[12]

The Minister for Federal Affairs and Nationalities Vyacheslav Mikhaylov explained the actions of the federal centre in Dagestan by stating, 'We treat the terrorists as terrorists need to be treated'. He went on to note that the attackers in Dagestan had their base in 'Chechnya' and that they were a multi-ethnic group that also contained a number of 'Arabs'. He concluded that:

> Chechnya has become an international terrorist base.... We are dealing with an attempt to realize, by the use of violent and terrorist methods, the extremist idea of a united Islamic state in Dagestan and Chechnya around which bandits, criminals and fanatics have gathered.[13]

Another clear feature of official Russian statements during the war in Dagestan was the argument that it was Russia that had been attacked, and not the other way around: Russia was acting in defence, and was not the aggressor in this war.[14] This argument was reinforced after a new attack in the Novolak region of Dagestan on 7 September 1999, when fighters took control of seven villages, even threatening the city of Khasavyurt. Speaking to the press on 8 September 1999, Putin declared: 'Russia is defending itself: we have been attacked. Therefore we need to throw off all our complexes, also our complex of guilt.'[15]

With these statements, Putin dismissed a core feature in the official de-securitizing narrative of the interwar period: that *Russia* was to blame for the problems in the North Caucasus. This created the rationale for a legitimate 'defensive' Russian counterattack. Following Putin's statement, the NTV channel, the very channel known for its 'pro-Chechen coverage' during the first war, broadcast horrifying scenes of a hostage pleading with the Russian authorities to defend those who were in 'Chechen slavery', before being decapitated with an axe by a masked fighter.[16] That footage underscored the inhuman nature of the enemy that Russia was facing, while identifying Russia as the righteous defender. The representation of Russia as victim and righteous defender was repeated by Putin immediately before the full ground offensive started in October: 'Today we are a victim of international terrorist aggression. This is no civil war.'[17]

The bomb blasts that hit Moscow, Buynaksk and Volgodonsk in late August and early September 1999 have been widely represented as the events that triggered the Second Chechen War. The first bomb exploded in a shopping centre near Red Square on 31 August. On 4 September, another bomb exploded in a block of flats in Buynaksk; on 9 and 13 September, two blocks of flats in Moscow were blown up; and on 16 September a bomb exploded near a block of flats in Volgodonsk. The perpetrators of these terrorist acts have never been identified. The Chechen President Aslan Maskhadov immediately distanced himself from these acts, as did the 'invaders' of Dagestan Emir Khattab and Shamil Basayev, and the FSB, who were also accused of complicity in these acts. Much has been said and written on this subject, but it still seems impossible to verify who committed these terrible acts. For our account here, of course, the main point is how they were represented. Regardless of who was responsible, these actions brought forth an abundance of statements that included representations of the threat and the threatened, and advice on the 'way out'.

After the second bomb blast in Moscow, Putin commented on 10 September 1999, 'in the course of Russia's history there have repeatedly been attempts to scare us and bring us to our knees, but nobody has managed. I have no doubts that no one will manage this time either.'[18] This representation of Russia as inevitably victorious was repeated by Putin several times later. When thanking the Russian (*rossiyane*) people for their response to the terrorist attacks he said 'no panic, no forbearance with bandits. That's the attitude for bringing the battle to a victorious conclusion, and we surely will.'[19]

Reinforcing the representation of 'Russia' under attack, heavily underscored by the new blast in Moscow on 13 September, President Yeltsin addressed the Russian people with the following words: 'Citizens of Russia, today is a day of mourning; a new disaster has befallen us. There has been another explosion with new victims. Yet another night-time explosion in Moscow. Terrorism has declared war on us – the Russian people.'[20] He went on to say:

> We are living amid a dangerous spread of terrorism that demands the unity of all forces in society and the state to repel this internal enemy. This enemy does not have a conscience, shows no sorrow and is without honour. It has no face, nationality or belief. Let me stress – no nationality, no belief.[21]

Putin, in televised remarks, characterized those behind the explosion as follows: 'It is difficult even to call them animals. If they are animals, then they are rabid.'[22] Yeltsin used a similar analogy in a televised address: 'They are trying to demoralize the authorities, to act covertly like wild beasts that sneak out at night to kill sleeping people without acknowledging their responsibility.'[23] Putin later used the animal analogy several times when characterizing Chechen fighters.[24]

On 13 September 1999, speaking at the opening of the autumn session of the Russian State Duma, Putin stated:

> this is not only about the conflict in Dagestan or the terrorist acts in Moscow; this is about protecting the security of the entire Russian statehood.... It's obvious that in Dagestan and in Moscow we are dealing with well-trained international terrorists, not with individual rebels. They are not laymen, but professionals specializing in subversive acts in the broadest sense of the word. Those who have organized and implemented the recent series of barbaric terrorist attacks are nursing far-reaching plans. They are trying to fan political tensions in Russia, and their main goal is to destabilize the situation in the country.[25]

In many statements – and there indeed were many of them – these themes were repeated.

In official statements the responses proposed for countering the threat facing Russia were either 'hard', 'tough', 'decisive', 'energetic' or 'uncompromising'; the terrorists needed to be 'annihilated' and 'destroyed'; any kind of soft approach would mean the destruction of Russia.[26] Prime Minister Putin's statement, given during a visit to Astana, which has often been translated as 'We will pursue them anywhere, and if, excuse the expression, we catch them in the lavatory we will waste them in the can',[27] became a much-cited phrase. This translation is somewhat misleading, however, as the expression 'Будем мочить их в сортире' is used in the jargon of the Russian criminal underworld as a very crude way of expressing 'to murder'.[28] Putin was indicating the kinds of methods necessary in the fight against this threat, to secure the survival of Russia.

The call for *unity* in withstanding the threat was another recurrent theme in official statements. This theme was introduced by Yeltsin in his initial response to the bombings, and was reiterated by Prime Minister Putin several times. In his speech to the State Duma on 14 September, Putin stated:

> having blown up the homes of our citizens, the bandits have blown up our state, they are not only attacking presidential power, the city's power nor the deputies power, but the country's power as such … terrorism has become a national problem.

He subsequently proposed unconditionally subordinating regional structures to the federal power ministries on this issue, urging lawmakers not to worry about issues of competency or authority, because 'people don't care who establishes order – the President, the Prime Minister, the *siloviki* or the Duma deputies'.[29] Over and over again, Putin made his appeal not only to the citizenry of Russia, but to the whole of Russian society, to all strata of power, be they 'the President, the Government or the Federation subjects', to act 'decisively, urgently and energetically'.[30] On 17 September 1999, during a debate in the Federation Council on the situation in the North Caucasus and the measures to be employed in fighting against terrorism, Putin called on the leaders of the regions to support the government, saying that the necessity of a hard fight against the bandit formations demanded the unity of all branches of power.[31]

Although the official argument was very clearly and consistently articulated in terms of the nature of the threat and the type of response required, questions remained as to where this response should actually be directed. Initially, both President Yeltsin and Prime Minster Putin took care not to name specific suspects in the Moscow bombings, but other political figures did not. After the first explosion in Moscow on 31 August, both the mayor of Moscow Yury Luzhkov and the Minister for Internal Affairs Vladimir Rushaylo immediately announced that they did not exclude the possibility of the terrorist act being connected to the situation in Dagestan.[32] Later, standing amid the ruins in Moscow after the 13 September blast,

Luzhkov said, 'we are naming Chechen bandits as the source of this terrorism'. Similarly, Rushaylo declared on NTV: 'What happened in Moscow was done by Khattab and Basayev and their people. There is no doubt about it.'[33] Moreover, the FSB had been warning for weeks that they had 'received operative information that diversionists from Chechnya were preparing terrorist acts in all major Russian cities'.[34]

Addressing the Russian State Duma the day after the final bomb explosion in Moscow, Prime Minister Putin dismissed any talk of linkages between the upcoming elections and the terrorist acts, branding such talk as 'open treachery, putting the authors of such speculation and provocation on a par with the terrorists'.[35] In his speech, which was the only televised part of the session, he spoke openly of the 'Chechen link' in the Moscow bombings. He did caution against practising repression on the basis of nationality, but also noted that Chechnya had become a 'huge terrorist camp'.[36] Such linking of 'Chechnya' to terrorism was repeated in later statements.[37]

A similar line of argument equating Chechnya with terrorism was made when Putin faced the Federation Council on 17 September. Prior to the debate, a documentary film *On atrocities committed by Chechen fighters* was screened, with terrible scenes showing how hostages were tortured and executed in Chechnya. Putin then delivered his speech, in which he said that the Chechen incursion into Dagestan had been carried out with the support of international terrorism. Using a language distinctly related to the discourse in the interwar FSB document referred to in Chapter 3, Putin stated:

> groups under the leadership of foreign masters are increasing their subversive activity … the interest in Chechnya from enemies in Muslim countries, and not just one Muslim country, is increasing … the directors of the terrorist war want to destroy Russia and create a pseudo-Muslim state with a military dictatorship and medieval order.[38]

The film and Putin's speech – the only parts of the session to be televised – were seen by millions of Russians across the Federation. A few days later, a similar film was shown on Russian TV2.[39] There were also several other visual representations that served to identify Chechnya as a threat. For instance, Putin appeared on the popular RTR television show 'Mirror' on 18 September with a bundle of fake US dollars allegedly stemming from Chechnya and a map showing where the 'Islamic extremists' had struck.[40]

Statements and information supplied by FSB Director Nikolay Patrushev substantiated the Prime Minister's argument that those responsible for the terrorist acts were in Chechnya, also underlining the international link. 'The terror acts in Moscow, Volgodonsk and Buynaksk were carried out by the same group, and some of its members are now hiding in Chechnya' stated Patrushev.[41] Further, the FSB claimed to have information that Shamil Basayev and Ibn al Khattab had been training groups in their camps to carry out terrorist acts, and that their people had recently been abroad to meet Osama bin Laden, who had pledged financial support. Bin Laden had also promised to send another group of well-trained fighters to Chechnya.

During a short visit to Rostov Oblast to discuss with the regional leadership how to deal with the consequences of the terrorist attack in Volgodonsk, Putin said to the journalists that it was widely known that the well-known terrorist Osama bin Laden had been to Chechnya several times; further, a variant of the 'fight against banditry' was currently being discussed with the Americans.[42] The argument of an enemy beyond Russia's borders, in alliance with Chechnya, was also presented by Yeltsin. In connection with the establishment of a strict cordon sanitaire around Chechnya in late September, he stressed 'the necessity of a 100 per cent guarantee that no mercenaries, accomplices or emissaries from countries far away, nor any weapons or ammunition, could enter the North Caucasus'.[43]

Putin also indicated the Maskhadov regime as consenting to terrorism, by stating: 'if the official authorities in Chechnya are not capable of controlling the situation and do not run to help the [federal] centre, it signifies that the current situation is one which suits them'.[44] Although Maskhadov was not labelled a 'terrorist', official statements during 1999 and 2000 pinpointed his identity as unreliable and potentially 'one of them'. Any talk of dialogue with Maskhadov was followed by 'if he shows that he is constructive and shows willingness to free his territory from international band-formations'.[45] On 1 October 1999, Maskhadov was discounted as the legitimately elected president of Chechnya when Putin announced that the Chechen parliament elected in 1996 was the 'only legal organ of power in Chechnya: the legitimacy of all other organs of power in Chechnya is conditional, because they were not elected according to the laws of the Russian Federation'.[46] Minister of Justice Yury Chayka held that 'leaders and members of organs of power established on Chechen territory in violation of Russian law will be prosecuted, also according to criminal law'.[47]

On the whole, a merging of very diverse groups and individuals on the Chechen side into one category of 'terrorists' was underway in official statements during the autumn of 1999. As Russian troops were entering Chechen territory in October, Putin stated that 'people are tired of bandits, they don't want to let terrorists into their villages, and we will help them'.[48] Such words effectively labelled any armed opposition to the advancing Russian troops as 'terrorist'.

Undertaking to employ 'the harshest possible measures' to fight the 'terrorist' threat inevitably pointed in the direction of starting a war of some kind against Chechnya, a war that few in Russia had felt ready for. The failure of the Russian forces during the First Chechen War and the heavy price paid in terms of young Russian lives, with no victory in return, were images still vivid in the minds of most Russians. Deputy Minister of the Interior Igor Zubov addressed this dilemma by noting that 'today it would hardly seem appropriate to conduct a military operation against the Chechen fighters, because we could incur great losses', but added, 'the federal forces are morally and militarily ready for action on Chechen territory'. According to the journalist reporting on the press conference, Zubov made his audience understand the necessity of a military operation by declaring that the fighters might plan to attack Ingushetiya and that the conquest of the Prigorodny region of Ingushetiya was a real threat. He did not exclude the possibility of a re-run of Budennovsk or other terrorist acts.[49] Thus, the argument was that the overwhelming threat facing Russia would have to be countered by a ground offensive against Chechnya, even though such a policy had been unacceptable until that time.[50]

Even after the Russian leadership had explicitly stated that Chechnya was part of the Russian Federation and it was clear to all that the new military campaign was not merely about taking out 'terrorists' on Chechen territory but about re-taking control over Chechnya as a whole, the rationale articulated was one of anti-terrorism. In Putin's words, 'our final aim is to destroy the terrorists and their bases in Chechnya'.[51] This rationale was argued consistently once the war got underway, also directly to the Russian soldiers in the field. As *NeGa* journalist Aleksandr Shaburkin reported from the military base in Stavropol on 22 October:

> his [Putin's] visit to Mozdok has convinced the Russian military that nobody will stop them in the fight against the bandits. And even if the details of the military operation ahead have not been conveyed, judging by the mood among the soldiers, the basic goal of the action has been confirmed – the annihilation of all terrorists on Chechen territory.[52]

The official securitizing narrative

Taken together, official statements on the brief war in Dagestan and on the terrorist bombings in Russian cities can be seen as the core of a 'securitizing narrative'. In the Russian official securitizing narrative of 1999, the threat was most frequently labelled 'bandit', 'terrorist', sometimes 'international terrorist'. Another frequently used epithet, particularly during the war in Dagestan, was 'Wahhabi'. The frequent and interchangeable use of the words 'Wahhabi' and 'terrorist' had important effects. The word 'Wahhabi', as noted earlier, need not always imply 'militant Islamist'. Very often and in other areas of the world it simply refers to a person who adheres to and lives according to the puritan Wahhabi code. The repeated linking and interchangeable use of 'terrorist' and 'Wahhabi' in Russian, however, has resulted in the construction of 'Wahhabi' as 'violent'. Here we should note that this puritan Muslim movement, which began to spread in Dagestan in the late 1980s, had already acquired quite a large following by that time. Adjustments to the meaning of 'Wahhabi' have served to attach a higher level of threat to this entire group.

A recurrent connection of 'terrorist' to 'Muslim' would have a similar effect. Indeed, a link between 'terrorist' and 'Muslim' was eventually made by Prime Minister Putin in his speech to the Federation Council referred above, but official language at this time seldom connected 'terrorist' to 'Muslim'. Of greater importance for this study is the constant linking of a third set of words: 'terrorist' and 'Chechnya'.

Although the terms 'Wahhabi' and sometimes 'Muslim' appeared together with 'terrorist' in official accounts, the words 'bandit' or 'illegally armed formations' – some of the most frequently used characterizations of Chechen fighters during the first war – were often used interchangeably with 'terrorist', making them appear as synonyms.[53] This merging of the previously dominant 'banditry' discourse on Chechnya with the emerging discourse on 'terrorism' created an indirect association between 'Chechnya' and 'terrorism' from summer 1999 onwards.

Moreover, the fact that both the incursion into Dagestan and the bomb explosions in Russian cities were labelled 'terrorist' and then eventually explicitly tied to 'Chechnya' in official language served to construct Chechnya as a 'terrorist' threat. Yeltsin's statement that 'we want to end, once and for all, the centre of international terrorism in Chechnya' epitomized this, and accompanied the new ground offensive against Chechnya on 27 October 1999.[54] As we will see in the chapters presenting representations by the Russian political elite, experts and media, equating 'Chechen' with 'terrorist' became widespread during autumn 1999. In turn, establishing Chechens as 'terrorists' became a social fact that made possible the exclusive and violent treatment of them as a group.

As for the representation of President Maskhadov as an 'event within the event', the meaning attached to 'Maskhadov' in official statements at this time did not equate him with 'terrorism', although there were changes in relation to the official interwar representations of Maskhadov. No longer represented as a victim, he was now portrayed as consenting to terrorism, by means of the expression that it 'suits him'. This linking of 'Maskhadov' to ideas of violence and guilt was re-emphasized when the Ministry for Internal Affairs spokesman on 17 September 1999 informed the press that they had evidence that a 3,000-strong unit of the Chechen army directly subordinated to President Aslan Maskhadov took part in the fighting in Dagestan.[55]

Additionally, three characteristics in the construction of the threat can be extracted from official statements on the incursion into Dagestan and the bombings in Russian cities. Together they serve to qualify the representation of the Chechen threat as *existential*, constructing an identity that can be placed at the top end of the scale of Otherness. The first characteristic is the *inhuman* nature of the threat. Many characterizations went far beyond words such as 'bandits', 'criminals', 'fanatics' or 'extremists', which all signify belonging to human society, albeit on its fringes. According to official statements, this particular enemy 'does not have a conscience, shows no sorrow, and is without honour',[56] making it 'difficult even to call them animals'.[57] (The reference to animals was repeated several times.) The inhuman nature of the enemy became even more frightening in relation to the deeds of which it was capable, represented not only in words such as 'barbaric', 'violent' and 'terrorist methods', but also in the stark audio-visual presentations.

Representations of the threat as inhuman were combined with a second set of descriptors portraying it not as erratic, irrational or inconsistent, but as 'professional' and 'well-trained', 'specialized' and with 'far-reaching plans', that further heightened the implied level of danger, particularly when combined with a third characterization of the threat as elusive, yet powerful – as expressed by terms such as 'wherever they might be', 'not individual rebels', 'dangerous spread' and 'huge terrorist camp'. The power of the threat was further amplified by references to a more distant but related enemy, as implied in expressions such as 'enemy circles in Muslim countries', 'the directors of the terrorist war' and 'Osama bin Laden'.

This threat had 'declared war' on Russia; it was attempting to 'bring us to our knees', potentially 'destabilizing the country' and threatening the 'entire Russian statehood'. Such expressions constructed the terrorist threat as a growing phenomenon that could engulf the entire country. This description of the kind of future Russia would face unless steps were taken implied that Russian authorities were standing at *the point of no return*. Response was urgent.

As noted, the possible *way out* suggested in this narrative was to initiate 'the toughest measures possible', or a 'hard', 'decisive', 'energetic' or 'uncompromising' response; the 'terrorists' needed to be 'annihilated', 'destroyed' or even 'wasted in the can'. A *united* Russia, particularly the unity of all branches of power, was presented as a precondition for succeeding in this struggle. Such rough policy suggestions were fully consistent with the identity construction that they drew upon. The nature of the threat, described in terms that served to dehumanize the terrorists, was such that it could not be dealt with by the use of law or common sense. Thus, prescriptions for a possible way out fitted the representation of the existential threat; they seemed both legitimate and necessary, given the future that Russia would face if such action were not taken. Within this discourse, we should note Putin's call, in a speech to the Duma in September 1999, for an 'unbiased analysis of the content and practices (*praktiki primeneniya*) of the Khasavyurt Accord', which in essence meant scrapping policies such as negotiation with Chechnya, or economic assistance to it.[58]

The securitizing narrative inherent in official statements also included a new representation of Russian identity. As argued in the theory chapter, it is difficult to imagine a re-articulation of the Other as a threat without a re-articulation of the Self. I would argue that adjustments made in representations of the Russian Self during the initial securitization of the Chechen threat in autumn 1999 marked the beginning of a radical re-articulation of Russian identity in official language – not necessarily in comparison to historical representations, but in comparison to dominant representations in official language since 1991, and particularly within the interwar discourse on Chechnya. This re-articulation continued throughout Putin's subsequent presidential terms (2000–2008), becoming even more explicit in his new term as president from 2012.

There were several aspects to this re-articulation at the early stage in 1999. First, in official statements, Russia was portrayed as the 'target of attack', under 'constant threat' and the object of a 'declared war'. Russia was not the offender: it was simply 'protecting' itself. Guilt could be placed squarely on the shoulders of the Other. Putin's 'we need to throw off all our complexes, also our complex of guilt' turned on its head the notion of Russia as the culprit, as articulated during the interwar years. This version of who was to blame was reinforced by Putin's account of the Khasavyurt Accord. In his speech to the Federal Council, he told the Duma that while Chechnya did not fulfil the provisions in the agreement, Russia had fulfilled all of them.[59] This idea of Russia's innocence has been strongly and consistently articulated in official language on the Second Chechen War.[60] Putin even represented the streams of thousands of refugees fleeing Chechnya during the heavy bombardment in early October 1999 as 'the Chechen nation voting with their feet against the criminal regime'.[61]

This construction of Russian identity in the face of the terrorist/Chechen threat carried with it a sense of revival and moral strength. Moral strength was paired with ideas of physical strength, in statements which referred to historical experiences stressing how 'Russia cannot be brought to its knees; nobody has succeeded before'. Putin turned to the past to deliver his fellow Russians from their inferiority complex. Another recurrent argument that contributed to the re-articulation of the Russian Self was that, in its response to the threat, Russia was bringing 'order' and 'discipline', thus representing Russia as the radical Other of the 'rabid', 'barbaric' and 'violent' terrorist/Chechen threat. Although official statements seldom drew boundaries between identities according to ethnic or religious lines (juxtaposing 'Chechen' or 'North Caucasian' or 'Muslim' with 'Russian' or 'Slav' or 'Christian'), it did happen. For example, official statements contributed to the emerging discourse on a 'genocide' being carried out against the *Slavic* population in the Northern Caucasus.[62]

Conclusion

The number of official securitizing statements in the period until Vladimir Putin became Prime Minister remained fairly low and contained an awkward and self-contradictory mix of representations that could be placed in the 'discourse of reconciliation' and representations that could be placed in a 'discourse of war' that constructed Chechnya as an existential terrorist threat. However, from early August, official statements brought a new intensity into the debate, not only in terms of how *many* official statements on the terrorist/Chechen threat could be found, but also in terms of how these statements contributed to strengthening the 'discourse of war' *alone*. There was hardly one official statement that could be counted under the 'discourse of reconciliation'. This was indeed an accumulation of official statements (and visual images) on 'terror' and 'Chechnya' that contributed to bring an urgent security situation into being. Even the controversial policy of another all-out war against Chechnya could be suggested under cover of the existential terrorist threat invoked in official representations.

The securitizing narrative offered in statements by the Russian leadership during autumn 1999 combined old and new ways of representing the Chechen threat. The narrative was clear and simple and fairly consistent across statements. Comparisons with official securitizing efforts during the First Chechen War (1994–1996) are instructive. Studies of Russian securitization of the Chechen threat before and during the First Chechen War have concluded that such efforts were few (in terms of how many official statements were given) and too late. A central argument had been that it was important to keep Chechnya in the Russian Federation because 'multi-culturalism' was a crucial feature of the Russian state character. Moreover, the historical community between the Russians and the Chechens was often referred to in official statements, while the Chechen population was made distinct from the Dudayev regime, portrayed instead as hostage to a small clique of leaders.[63] Any attempts to demonize the Chechen opponent in official statements thus implied an internal inconsistency in the narrative, as such attempts contradicted the claim that the Chechens were not so different from the Russians.

While official statements during the First Chechen War underlined positive identification with the group on whom war was to be waged, the Second Chechen War was launched to the accompaniment of statements that constructed a consistently negative, one-dimensional and indeed frightening image of the terrorist/Chechen threat. According to the 1999 official narrative, Russia was facing an existential threat from one internal enemy – 'terrorism' – which encompassed all factions of the Chechen separatist movement, albeit some involuntarily, and working in alliance with a distant enemy. Moreover, the inhuman nature of this enemy and the fact that it had already struck in Dagestan as well as in the heart of Russia called for immediate emergency action against Chechnya in order to secure the very survival of Russia. The possible 'way out' indicated in the narrative not only scrapped practices such as negotiations and economic relief as means of dealing with Chechnya, but urged the destruction of the threat by any means. Chechnya had become a question of survival for Russia, a *casus belli* over which blood would have to be shed.[64]

At the same time, this extremely threatening re-articulation of the Chechen threat generated a re-articulation of Russian identity that broke with the humbler version predominant during the interwar period, and now projected Russia as innocent, strong and capable of establishing order. Thus, through the official securitizing statements launched during summer and autumn 1999, the Russian leadership promised security to the people, but also re-defined Russian identity. Chechen independence, as a threat to Russian territorial integrity, was not securitized at all in official language. Chechnya as a part of Russia, the issue over which the First Chechen War had been fought, was simply stated as a self-evident fact in short phrases, such as 'I repeat – Chechnya is Russian territory, and we can place our forces where it suits us'[65] and 'there is no border with Chechnya'.[66] Official statements settled this crucial issue from the very beginning – not explicitly, but under cover of a securitizing narrative which portrayed Chechnya as an existential terrorist threat.

As to the wider implications of this new official 'securitizing narrative' for the object that was securitized, the most important point lies outside the 'narrative' itself, and concerns the linking of the 'terrorist threat' to 'Chechnya'. I hold that the sum of official statements created an equation between these two objects. This happened through the re-circulation of the descriptor 'bandit' (the old term widely applied to Chechen fighters) together with and as a synonym for 'terrorist', as well as through the explicit references to 'Chechnya' as being 'terrorist'. Moving on from 'Chechnya' as a territory to the population of this territory, in official statements the legally elected Chechen President 'Maskhadov' shifted from being an ally of Russia (himself threatened by the extremists) to being 'their' ally, threatening Russia.

Even more problematic, perhaps, is the fact that official statements did *not* distinguish the Chechen civilian population from the 'terrorist' threat in any explicit way. With no explicit positive identity attached to this group, it was easily subsumed under the terrorist label as well. The consistent and many-layered securitizing narrative outlining and detailing this terrorist threat as an *existential* threat to Russia and the violent policies and practices needed for dealing with it readily translated into an understanding of who the *Chechens* are and what *we* can and should do to them. As we will see in Chapters 10, 11 and 12, such a reconstruction of the Chechens would have grave implications for how this group of people could be treated once the war had begun.

Before moving that far, however, let me take a step back, and stress that the official statements reviewed in this chapter should be considered only as an evolving 'securitizing *move*' in the terminology of securitization theory. A securitizing move does not automatically translate into the securitization of an issue and the thereby-legitimated undertaking of emergency measures. How the audience receives the securitizing move will depend on the discursive terrain into which it is launched. Crucially, the audience also gets a say: the securitizing narrative can be confirmed, but also re-written or negated. The official narrative extracted and detailed in the present chapter will serve as a basis for assessing how well the official narrative resonated with the historical discursive terrain in Russia, as well as how it was received and negotiated in the putative 'audiences' during autumn 1999.

Notes

1 N. Gevorkjan, N. Timakova and A. Kolesnikov. 2000. *First Person: An astonishingly frank self-portrait by Russia's president.* New York: Public Affairs, 139.
2 'Boris Yeltsin upovayet na silovikov', *NeGa*, 17 August 1999.
3 'Duma dala Vladimiru Putinu neobkhodimoye', *NeGa*, 17 August 1999.
4 'Kompleks mer po navedeniyu poryadka', *RoGa*, 11 August 1999.
5 'Mezhdu Andropovym i Pinochetom', *Kommersant*, 11 August 1999.
6 'I.O. Prem'yera dolzhen podtverdit' svoyu reputatsiyu', *NeGa*, 11 August 1999.
7 'Reshitel'no, no ostorozhno', *NeGa*, 12 August 1999.
8 In these villages a 'pure version' of Islam was practised and they had been declared 'independent Islamic territory' one year earlier.
9 'V Dagestane prodolzhayetsya silovaya operatsiya', *NeGa*, 31 August 1999.
10 'Informatsionnaya blokada konflikta v Dagestane', *NeGa*, 14 August 1999.
11 'Moskva i dal'she budet podderzhivat' opolchentsev', *NeGa*, 31 August 1999.
12 'V Dagestane prodolzhayetsya silovaya operatsiya', *NeGa*, 31 August 1999.
13 'O normalizatsii obstanovki v Dagestane', *NeGa*, 24 August 1999.
14 Putin noted in connection with a planned visit to Dagestan by Yeltsin that 'the Dag-estani nation needed moral support because they had risen with guns in their arms to defend their own homes' ('Pered reshayushchim shturmom', *NeGa*, 18 August 1999).

15 'Kreml' izbavlyayetsya ot kompleksa viny', *NeGa*, 9 September 1999.
16 Ibid.
17 'PM Putin vows to Stay the Course', *Moscow News*, 29 September 1999.
18 'Ponedel'nik v Rossii ob''yavlen dhem traura', *NeGa*, 11 September 1999.
19 'Na Lubyanke znayut, kto sovershil terakti', *NeGa*, 25 September 1999.
20 'Text of Yeltsin address on Moscow bombings', *Reuters*, Moscow, 13 September 1999. A similar version of the Russian people as being under attack by terrorism was repeated in a telegram from Yeltsin to the governor of Rostov after the last bomb exploded in Volgodonsk ('Nado zadushit' gadinu na kornyu', *RoGa*, 17 September 1999).
21 'Text of Yeltsin address on Moscow bombings', *Reuters*, Moscow, 13 September 1999.
22 'Moscow awash in explosion theories', *Moscow Times*, 14 September 1999.
23 Ibid.
24 For example, on live television when the Chechen warlord Salman Raduyev was captured ('Terrorists are people, not animals', *Moscow Times*, 21 March 2000).
25 'Putin. No need to pass new law on state of emergency', *Interfax*, 14 September, as carried on Johnson's Russia List, 15 September 1999; and 'Putin predlagayet novyy plan Chechenskogo uregulirovaniya', *NeGa*, 15 September 1999.
26 Putin's statements, reported in 'Nado zadushit' gadinu na kornyu', *RoGa*, 17 September 1999; see also his lengthy interview with editors of regional newspapers in 'Vladimir Putin: 'Chechnya zanimayet tol'ko 45% vremeni v rabote pravitelstva', *Chas Pik*, 20 September 1999; or Defence Minister Igor Sergeyev, quoted in 'V Kreml' cherez Chechnyu?', *Segodnya*, 28 September 1999.
27 'Voyna bez vykhodnykh', *NeGa*, 25 September 1999.
28 I am indebted to Russian philologist Maria Kim for pointing this out.
29 'Putin predlagayet novyy plan chechnskogo uregulirovaniya', *NeGa*, 15 September 1999.
30 'Hado zadushit' gadinu na kornyu', *RoGa*, 17 September 1999. A similar call for consolidation of power as the only means of fighting the terrorist threat facing Russia was reiterated by Putin during a government meeting on 16 September (Buynaksk, dva raza Moskva', *NeGa*, 17 September 1999.
31 'Khasavyurt byl oshibkoy', *Trud*, 18 September 1999.
32 'Terroristicheskiy akt v tsentre stolitsy', *NeGa*, 2 September 1999.
33 'Moscow awash in explosion theories', *Moscow Times*, 14 September 1999.
34 'Seyat' uzhas i smer't v rossiyskikh gorodakh', *Kommersant*, 11 August 1999.
35 'Tret'ya otechestvennaya?', *Monitor*, 15 September 1999.
36 Ibid.
37 On 15 September, Putin again accused Chechnya of providing refuge for the perpetrators of the Moscow apartment building blasts, who he said were receiving support from 'Chechen extremist forces' ('Vzryvaya doma, bandity vzryvayut gosudarstvo', *RoGa*, 16 September 1999).
38 'Ot truslivogo zaytsa', *Obshchaya Gazeta*, 23 September 1999. See also statements from Putin's press conference on 27 September reported by *ITAR-TASS*, 27 September 1999.
39 Jeffrey Tyler described how:

> Last weekend Russian television's Channel 2 showed grisly video footage purportedly shot by Chechen rebels: A bearded and swarthy Chechen guerrilla is kneeling on the back of a blond, tied-up and panicked Russian soldier of 18 or 19. The rebel takes a foot-long knife and, smiling at the camera, methodically saws through the squirming boy's throat and neck, bloodily working the knife back and forth, until his head comes off. The guerrilla holds the severed head up for the camera and laughs.
>
> ('Russia on the Edge', 2 October 1999, available at www.salon.com/travel/ feature/1999/10/02/moscow, and accessed 27 February 2016)

40 'Telebitva za golosa izbirateley nachalas'', *NeGa*, 21 September 1999.
41 'Vikhr'-antiterror dayet polozhitel'nyye rezul'taty', *NeGa*, 29 September 1999.
42 'Vladimir Putin posetil Don', *NeGa*, 24 September 1999.
43 'Vstrechi v Kremle', *NeGa*, 28 September 1999.
44 'Ob''yavlen karantin', *Vremya MN*, 15 September 1999. The argument that Maskhadov became one of the terrorists when he failed to distance himself clearly from them was also a recurrent argument later on. For example, when Putin rejected Maskhadov's proposal to negotiate in October 1999, he also said that Maskhadov had painted himself into a corner by 'establishing social relations with people considered international terrorists' ('Putin ne poveril Maskhadovu', *NeGa*, 12 October 1999). Speaking on television on 20 November 1999, Putin vowed that the campaign would continue until all 'terrorist bands' were eliminated, ruling out any negotiations with what he termed 'international terrorists' (*FRE/RL Newsline*, 22 November 1999).
45 Putin, reported in 'Magomedov ne vtretilsya s Maskhadovym', *NeGa*, 30 September 1999.
46 'Nevernyy shag v pravil'nom napravlenii', *NeGa*, 2 October 1999.
47 'Putin sozdal klub prem'yer-ministrov', *NeGa*, 6 October 1999.
48 Ibid.
49 'K novoy voyne v Chechne pochti vse gotovo', *NeGa*, 23 September 1999. Putin also dismissed a ground offensive in late September, saying that:

> no broad military operation in Chechnya was planned.... Our task is to save the Russian population from the bandits, but in what way you will soon discover. It will be nothing of the kind that was during the so-called sad famous Chechen campaign.
>
> (Ibid.)

50 A similar line of argument was given by other government representatives over time: in mid-October, Ramazan Abdulatipov argued: 'As long as the bandits are not totally destroyed, there will not be peace for the Chechen nation nor for nations in other Russian regions ... in this situation there is only one way out – to destroy them'. He did not exclude the possibility of official talks with Maskhadov:

> if he from his side would distance himself from the terrorists and not undertake a common fight with them, but this is not happening ... there will definitely be terrorist acts, and brutal acts against Russian forces that go into Chechnya.
>
> ('Bandity podlezhat unichtozheniyu', *RoGa*, 16 October 1999)

51 'Novaya granitsa Ichkerii', *NeGa*, 6 October 1999.
52 'Grozny budut brat' po chastyam', *NeGa*, 22 October 1999.
53 See the quotes referred to in 'Defining the threat' above: for example Putin's comment that 'people are tired of bandits, they don't want to let terrorists into their villages and we will help them' ('Putin sozdal klub prem'yer-ministrov', *NeGa*, 6 October 1999).
54 'Vokrug Groznogo szhimayetsya kol'tso', *NeGa*, 28 October 1999.
55 'Interior Ministry claims proof of official Chechen involvement in Dagestan fighting', *RFE/RL Newsline*, 20 September 1999.
56 'Text of Yeltsin address on Moscow bombings', *Reuters*, Moscow, 13 September 1999.
57 Putin in televised remarks, referred to in 'Moscow awash in explosion theories', *Moscow Times*, 14 September 1999.
58 'Karantin dlya virusa voyny', *Vek*, 19 September 1999.
59 'Sovet Federatsii podderzhivayet zhestkiye mery', *Russkaya Mysl'*, 23 September 1999.
60 Official rhetoric has constantly rejected the idea of a 'humanitarian catastrophe' as a consequence of war; members of the government have kept insisting that Russia is handling everything very well (see, for example, Minister of Emergency Situations, Sergey Shoygu 'Yest' lyudi, sposobnyye na provokatsii', *Segodnya*, 12 November

1999; or 'Ya nikogda ne stanu volkom seroy stai', *Komsomol'skaya Pravda*, 1 October 1999).

61 'Nas tak zashchishchayut, chto dazhe stydno', *Novaya Gazeta*, 4 October 1999.

62 For example, head of the temporary information centre of the Ministry of Defence Andrei Matviyenko said that 'Slavs, in particular Russians are being subjected to slaughter, theft and violence' ('Dva vzglyada na problemu bezhentsev', *NeGa*, 7 October 1999).

63 C. Wagnsson. 2000. *Russian Political Language and Public Ppinion on the West, NATO and Chechnya.* Stockholm: Akademitryck AB Edsbruk, Ch. 5.

64 I.B. Neumann. 1998. Identity and the outbreak of war. *International Journal of Peace Studies*, 3(1): 7–22.

65 Putin to the CIS Council of Heads of Secret Services ('Patrushev vozglavil sovet rukovoditeley spetssluzhb SNG', *NeGa*, 1 October 1999).

66 Putin, in an interview at the airport in St Petersburg ('Rossiya ne schitayet metry', *NeGa*, 1 October, 1999).

6 Historical representations of Chechnya and Russia

Securitizations can happen quickly. The time that passes from an emerging accumulation of securitizing statements to the multiplication of such representations of existential threat in the wider public resulting in a situation of 'consent in the audience' is not necessarily very long. This swiftness is logical, given the urgency that is constructed in most securitizing narratives, but it cannot be taken for granted. It is conditioned upon the consistency of the narrative and how well it is argued, but also upon the discursive terrain, the specific cultural context in which the securitization takes place.

In the case of Russia in autumn 1999, the official securitizing narrative was built up and presented to the Russian public in the course of less than a few months before the most radical emergency measure – war – was undertaken. As explained in the previous chapter, the securitizing narrative put forward by the Russian leadership was not only frightening, but also consistent, simple, one-sided and supported by strong visual images. The security argument itself can be said to have been convincing. And, as we will see, consent emerged fairly quickly (Chapters 7, 8 and 9). The Russian public, broadly speaking, had accepted new representations of Chechnya as a *casus belli* over which blood would have to run as the tanks rolled into Chechen territory.

However, according to the securitization theory elaborated in this book, securitization is not limited to the articulation of a convincing security claim by a few: it is produced in a complex intertextual relationship and draws on previous representations of existential threat as well as those constantly emerging among groups in the public that are construed as 'audience', but which are actually also speakers. As suggested in the theory chapter, 'securitizing moves' are not launched into empty discursive space, but are structured by and resonate with latent or manifest representations in pre-existing discourses. Existing discourses privilege and disadvantage certain securitizing moves, as opposed to others. A securitizing narrative that resonates well with and draws on recurrent common meanings and identity constructions in the national discursive terrain will acquire legitimacy through this resonance.

Representations of Chechnya and Russia, implicit in the official securitizing narrative and presented in Chapter 5, did constitute a break with the dominant official discourse of the interwar period. But that does not mean that there had not been fertile ground for these representations in historical representations or in more current, non-official representations of Chechnya and Russia. We have already seen how the discourse of war was strong in Russian media representations as well as in those of the FSB during the inter-war period (1996–1999). This chapter presents a broader outline of discourses on Chechnya and Russia prior to 1996. I cast the net much wider in time and space to uncover the discursive terrain which the 1999 securitizing move was launched into and which it fed on and resonated with.

The account of historical debates on Chechnya and Russia starts by reviewing classical Russian literary representations in the nineteenth century. I then sketch the representations that dominated official discourse in the Soviet period as well as those that accompanied the First post-Soviet Chechen War, official as well as media representations. Finally, representations found in texts of the nationalist and communist opposition prior to 1999 are presented. The account draws largely on secondary literature. Several scholars have already investigated articulations and re-articulations of Chechnya as one of Russia's habitual Others. The sum of these articulations is taken as a sounding-board for the official securitization of the Chechen threat in 1999. Throughout this chapter, the official narrative extracted and presented in Chapter 5 will be compared to dominant representations of the relation between Chechnya and Russia found in key texts pre-dating 1999.

Tsarist and Soviet era representations

Russian literary representations of the Northern Caucasus can be traced back to the Russian poets of the early nineteenth century. At that time, violent encounters between the Russian empire and the peoples living in this region were well underway. As there was no reportage from the frontline, these literary accounts acquired high importance. Indeed, according to Harsha Ram 'Russia's literary tradition was the primary locus of Russian debate on the Caucasus until the media revolution of the post-soviet nineties.'[1]

While the poetry of Mikhail Lermontov vacillated between demonizing and ennobling the peoples of the Northern Caucasus, the simplest and most potent myth was that of a Wild Man who posed a constant violent threat to all that was Russian. Ram also indicates that Chechen society was naturalized as a savage one in Russian poetry, in which war and freedom were the most dominant features – a kind of anarchy. Yet, this literature also acknowledged that customary law and blood feud served as well-functioning codes of behaviour. Moreover, the myth of the Caucasian as a Savage engaged in perpetual warfare also had a counterpart in the imagery of the *Noble* Savage. Thus, what we find in representations of relations between Russia and the North Caucasus in this literary discourse is no rigid hierarchy, but an ambiguity combining fear and admiration.

Representations of the Russian state were similarly ambivalent. The Russian as a 'Prisoner of the Caucasus' was a recurrent theme, but within this imagery the Russian was also represented as a prisoner of the Russian autocratic regime.[2] Russian classical poetry thus levelled a criticism against the imperial regime, arguing that, in seeking to subjugate the Caucasus, Russia had become its prisoner. On the whole, the nineteenth-century literary discourse on the Caucasus indicated a civilizational divide between the Savage and the Colonizer, but it also transmitted cultural empathy and pointed to divisions *within* Russia as well.[3] Rather than a 'Radical Other', the literary representations projected what Croft refers to as a 'Orientalized Other', 'one in which the Self is deeply engaged, sometimes attracted, sometimes frustrated'.[4]

Studies of official/public language on the Caucasus/Chechnya are not easy to find, but the ambivalence found in classical literary representations probably did not exist in the language of politicians and military men of the time. In his well-documented history of Caucasian nations, Oliver Bullock gives several references to military discourse from the period of Russian colonization of the Caucasus. Here the North Caucasians are referred to as 'rogues' and 'rascals', in fact hardly proper humans at all.[5] Moving into the Soviet period, official documents show that branding these people as 'bandits' was widespread.[6] Even the categorization of Chechens as 'terrorists' seems to go far back. In the plan for deporting some 450,000 Chechens and Ingush scheduled for 23 February 1944, the charges included 'active and almost universal involvement in terrorist activities directed against the Soviets and the Red Army'.[7]

This one-sidedly negative construction of Caucasians found a more systematic articulation in the 1960s and 1970s, when some Russian intellectuals began positioning non-Slavs in the Soviet Union – primarily those in the Caucasus and Central Asia – as the Other against which Russian national identity was formed, instead of the West, as previously.[8] In the Gorbachev period, these ideas became known among the wider public through the numerous popular periodicals and books. They were also articulated by emerging racist groups in the post-communist period. According to Vera Tolz, these groups 'view Central Asians and the Muslim Peoples of the North Caucasus, rather than the Jews as posing the greatest danger to the survival of Russians'.[9] In the early 1990s, the leader of the National Republican Party of Russia N. N. Lysenko proposed that all Muslims from the Caucasus and Central Asia be deported from the Russian state, and that Russians should be compensated for the economic genocide they had suffered at the hands of the southern mafia.[10]

Representations during the First Chechen War

'Criminality' emerged as one of the defining images of the Chechen diaspora in the 1990s; this imagery was adopted by the Yeltsin regime.[11] Ram indicates that, in official rhetoric, the Chechen was updated from a pre-national Savage, to a post-national Criminal, representing part of a contemporary transnational circuit of financial interests involving oil, drugs and weapons. Chechen criminality was also represented as spatially omnipresent, both within and beyond their borders.[12] A quote from Yeltsin's annual speech to the Federal Assembly is illustrative:

The organic fusion of the criminal world with political power – which both politicians and journalists have been speaking of incessantly as the main danger facing Russia – has become reality in Chechnya. It has been the launching pad for the preparation and diffusion of criminal power into other regions of Russia.[13]

Another recurrent term in descriptions of the Chechen adversary was that of the wolf. Again, the notion of 'wolf' is ambiguous and not necessarily negative.[14] In Russian imagery, however, it was negative only. Russell writes:

> perceived to be a fearsome, cunning, fierce but un-tameable opponent, for the Russians the wolf came to symbolize the Chechen, a worthy enemy, but one that was wild and dangerous enough to warrant only destruction. Lupine epithets were given to the Chechen leaders: Aslan Maskhadov – 'the wolf with a human face', Shamil Basayev – 'the lone wolf' and Salman Raduyev – 'the loony wolf'.[15]

In official discourse from the First Chechen War, the idea of Russia as a prisoner of the Caucasus was re-circulated, but not in this classical double sense. Rather, the official narrative portrayed Russia as the only victim, the Russian state largely benign in intention. At the same time, Ram notes:

> one is struck by how willingly the authorities here concede the porosity and anarchy of Post-Soviet space, and hence their own impotence as a centralizing force. The confident rhetoric of imperial expansion or socialist construction has been replaced in the 1990s by one of national emergency.[16]

In official representations, the state was like 'an increasingly passive witness to the wider shifts in the nation's political economy'.[17] Alla Kassianova draws similar conclusions on official articulations of Russian state identity in the 1990s in her analysis of key official texts such as the Foreign Policy Concept of 1993 and the National Security Concept of 1997. Russia is represented as crisis-ridden and weak, dependent upon support from the leading democratic states, or as being deficient.[18]

Dramatic changes in the Russian information sphere after the fall of the Soviet regime created entirely new possibilities for alternative articulations of Chechnya and Russia. This was a diametrically opposite development of the media sphere compared to the development from 2000 onward.[19] In the years following the collapse of the Soviet Union, the information field was open and a range of discourses challenging official representations prevailed during the First Chechen War. These also drew on the older and well-established Russian discourse on the Caucasus, but offered very different re-articulations of relations between Russia and Chechnya from official representations, particularly as the war unfolded.

Russian television, the printed media and the western documentary journalists all focused consistently on the same images: the corpses of civilians killed by aerial bombardment, the decomposing bodies of Russian soldiers abandoned by their own army to scavenging dogs, anxious Russian women travelling to the Caucasus in search of their missing sons and husbands conscripted into war and the hostage crises in Budennovsk that transfixed the nation for several days.[20]

These representations contained the criticism inherent in the classic notion of the Prisoner of the Caucasus: Russian civilian and military casualties were represented as victims of Russian coercion; the Russian nation had become captive to the regime's policy.

Media representations also blurred the sharp distinction between Self and Other indicated in official imagery by representing Chechens and Russians alike as victims of a senseless war. In Ram's words, 'what emerged was a spectacle of general carnage in which no distinction was made between rebel militias and an ethnically mixed local population'.[21] In the story told during the single largest terrorist act of the century, when at least 1200 hostages were held captive by Chechen fighters in a hospital in Budennovsk, the Russian soldiers came across as the brutal actors. Ram argues that even the leader of that mission, the Chechen warlord Shamil Basayev:

> readily embodied the Noble Savage ... feared to be sure, but nonetheless admired for his audacity; an outlaw, but one sympathetically viewed by many of the hostages themselves, who seemed more dismayed by 'their' government's response than by the actions of their captors.[22]

In the political sphere also, alternative discourses co-existed with official representations of the war. The media representations referred to above were not very different from the representations offered by, for example, Russia's Commissioner for Human Rights Sergey Kovalev who played a key role in Russian politics at the time. He denounced the gross violations of human rights and humanitarian law that occurred, and publicized the human cost of the war in Chechnya. Kovalev himself led a five-man group including several State Duma deputies to Chechnya to monitor the war there.[23] While in Grozny, the group relayed a series of bulletins and appeals on the war back to Moscow.[24] These reports highlighted civilian and military casualties, presenting them as victims rather than agents of Russian coercion. Although Russian casualties were the main focus, Chechens were shown as fellow victims, thus contradicting official representations constructing Chechnya as a radical Other. The strongest feature of these texts was their explicit criticism of the Russian authorities. The benign intentions ascribed to the Russian state in official rhetoric were replaced by notions of lies, lawlessness and cruelty in this narrative. The war that the Russian authorities were waging against Chechnya was depicted as gruesome and futile, a catastrophe and disgrace for Russia. To this it should be added that international society had access to Chechnya during the First War and thus could articulate their version of the war with credibility. Accounts from OSCE missions largely tallied with the alternative discourse prevalent in the Russian press at the time.[25]

Let us return to official representations during the interwar period (1996–1999) discussed in Chapter 4. The 'discourse of reconciliation' resonated with, and built on, these alternative representations of 'Chechnya' that emerged during the First War. Representations of 'Chechnya' and 'Russia' in the official narrative of 1999 (discussed in Chapter 5) contrasted radically with them. However, the 1999 official securitizing narrative was no invention: it was a re-articulation which combined old and not-so-old representations of 'Chechnya' as radical Other and new ones that were somehow a logical extension of these. Presenting an image of the Chechen terrorist threat as brutal, violent and gruesome, the 1999 narrative resonated with that portion of the ambiguous classical literary tradition which projected the North Caucasian as a 'Wild Man' who posed a constant violent threat to Russia. Indeed, Yeltsin's September 1999 warning that the terrorists are 'like wild beasts who sneak out at night to kill sleeping people' parallels Lermontov's 'Do not sleep, Cossack, in the darkness of the night; Chechens are moving beyond the river!'[26] This also reminds us that the imagery of Chechens as animals, which appears in the Yeltsin quote and which was a recurrent feature in the 1999 official narrative, was by no means new to the Russian audience. It was a well-established part of the Russian discursive terrain.

Moreover, the articulation of Chechens as terrorists was only one step removed from ideas in circulation in the 1990s which framed the Chechens as criminals. They could easily be reinserted into the debate. If the Chechen was updated from a pre-national Savage to a post-national Criminal in official rhetoric during the First Chechen War, as Ram indicates, it was updated again in 1999 according to the same logic, by adding 'terrorist' to the imagery. Just as Chechen criminality had been represented as part of a highly contemporary transnational circuit and as spatially omnipresent, both within and beyond Russian borders, during the First War, so was Chechen terrorism before the Second War was launched. Chechnya was identified as a node in the growing international terrorist network, elusive yet omnipresent and linked to enemy circles abroad. Thus, Yeltsin's October 1999 dictum – 'We want to end once and for all the centre of international terrorism in Chechnya' – can be seen as a logical extension of the argument for war given in 1995: 'The organic fusion of the criminal world with political power ... has become reality in Chechnya.'

As to what kind of match there was between representations of Russia within the official 1999 narrative and previous articulations, there can be little doubt that the 1999 narrative contrasted with official representations of 'Russia' during the First Chechen War. Official representations in the 1990s portrayed Russia as weak and deficient, and the media pronounced the guilt of the incumbent regime. However, other articulations of Russia are also relevant for understanding the discursive terrain that formed the backdrop to the 1999 official narrative. These alternative articulations of Russia were similar to and drew on the confident discourse of Russian imperialist expansion and Soviet construction. Iver Neumann (1996) identifies them as the re-emergence of the centuries-old Romantic nationalist position in the debate about Russia and Europe.[27] In the early 1990s they were promoted by the 'national patriotic bloc' – the nationalist and the communist opposition. It is particularly important to investigate these representations because, as Tolz points out, 'the opposition devoted much more attention to what Russia was than did the Yeltsin government'.[28]

Communist and nationalist representations in the interwar period

The discourse of the 'national patriotic bloc' expressed an optimistic view of Russia and its future, initially even conveying belief in the re-creation of the Soviet Union/empire.[29] Russia's uniqueness, greatness and potential strength was a recurrent theme in the texts of both nationalists and communists. Even if the West functioned as the radical Other in these texts, their articulation of Russia should still be considered when trying to map the discursive terrain that formed the backdrop to the 1999 official narrative. The account below does not in any sense present the full range of ideas within communist and nationalist texts, but focuses on the elements of relevance as a sounding board for the 1999 official securitizing narrative.

According to Luke March, Gennady Zyuganov (who had headed the Communist Party of the Russian Federation (CPRF) since 1993) increasingly made the statist patriotic orientation the cornerstone of the Party.[30] Zyuganov's texts drew on a range of different historical and often contradictory ideas and theories on what Russia was, but relied heavily on the 'Eurasian Idea' which can be traced back to nineteenth-century Slavophilism.[31] In his texts, Zyuganov represents the history of Russia as a constant struggle to secure its natural hegemonic position as a Eurasian power. Not surprisingly, the Soviet era is nostalgically represented as a positive period that provided Russians with international respect and pride in their country's achievements. As for the future, Zyuganov's writings suggest that Russia could be strong enough to stand up to the West only if it is a Eurasian power. As Smith writes, in Zyuganov's view 'Russia's geopolitical mission is to connect up historically with the idea of Russia as a Great power (Derzhava)'.[32] Zyuganov also argues that great-power status can be secured by a strong state with a strict and prudent authoritarian leadership.[33] At the same time, the idea of an organic link between party/state and the people figures strongly in Zyuganov's writings. Unity is secured by giving priority to common and collectivist interests over private, egoistic and individual ones.[34]

Zyuganov's Russia was thus an inversion of the 'weak' and 'subservient' Russia of the 1990s which was destined to 'disappear'.[35] Russia was predestined to 'show to the world the treasures of the human spirit, as embodied in her personal and family way of life, her social structure, and her great power statehood'.[36] Summing up the message inherent in these texts, Urban and Solovei write that 'Ziuganov invoked a Manichean picture of the world in which the centre of goodness, light, of all conceivable and inconceivable virtues – Russia! – was counter-posed to the pole of evil – the West.'[37]

The 'New Right' in Russia also drew heavily on the Eurasian Idea, emphasizing Russia's special position as part of a distinctive Eurasian civilization. The representation of Eurasia as inseparable from Russia's renewal and dignity served to underpin their argument that Russia needed to re-secure control over Eurasia and re-establish Russia's hegemonic geopolitical position towards the South.[38] While the principal adversary was the West, the New Right saw the cultural threat to Eurasianism as much broader. *Mondialism* – shorthand for globalization, cosmopolitanism and both liberal and socialist internationalism – was held to emanate from Western-based practices of 'chauvinistic cosmopolitanism'. 'As part of a carefully orchestrated and on-going subversive strategy to undermine Eurasianism and further weaken Russia, it is claimed that mondialism also had its "fifth columnists" within Russia itself.'[39] In an interesting twist, Aleksandr Prokhanov argued that Atlanticism had long attempted (unsuccessfully) to promote Islam as a buttress against Russia fulfilling its Eurasianist mission.[40]

As in the communist texts, the solution for Russia was the strong state and imposition of 'authoritarianism, which will make it possible to begin to stabilise chaos, blood and insanity, and then, through strong authoritarian power, the cultivation of democracy will slowly begin, not through the creation of insane parliaments, but corporative democratism'.[41] It should be noted that the notion of *unity*, as opposed to the disintegration and chaos associated with Yeltsin's Russia, was also articulated as an ideal by more moderate nationalist forces. In urging the unity of the Russian nation, the Congress of Russian Communities (KRO)[42] referred to the territorial unification of national territory and compatriots abroad (Russians outside the Russian Federation) but also called for social unity within the nation.[43]

In the political arena, the most prominent spokesman of New Right ideas was Vladimir Zhirinovsky, the leader of the electorally successful far right political party in Russia, the Liberal Democratic Party of Russia. The two main principles in Zhirinovsky's erratic and inconsistent body of texts are the primacy of the *russkiy narod* and the expansion of Russia as an Empire. However, Ingram writes that, according to Zhirinovsky, 'the state vies with the nation as the ultimate value in politics, but it is the state ... which is to take the active role in Russian development'.[44] As in Zyuganov's texts, Zhirinovsky draws inspiration from the need to redress past defeats and humiliations: 'We have suffered enough. We should make other people suffer.'[45]

Even if the West functions as the radical Other in these texts, Zhirinovsky's writings also construct 'the South' as the Other side in a civilizational divide. For Zhirinovsky, the importance of Russia's southward expansion had a positive side effect:

In the process, Russia can provide stability and order amongst the 'southerners', whose clan-based social structures are interpreted as the enduring cultural markers that distinguish Russians from the Eurasian South, and whose very social condition has a tendency to encourage organized crime, social disorder and ethnic conflict.[46]

The intention with this detour has not been to provide a full overview of alternative representations of 'Russia' in the 1990s, but to show that the articulation of Russian identity implicit in the 1999 official narrative was by no means alien to the Russian audience. Articulations of the Russian Self emphasizing strength and uniqueness, stripped of any notion of guilt, have enjoyed a constant presence in debates on Russia, historically and throughout the 1990s as well. Judging from the large numbers of votes cast for nationalists and communists in the 1993 and 1995 State Duma elections, such an articulation of Russian identity found strong resonance amongst the Russian population.[47]

While the West was usually projected as the radical Other in the language of nationalists and communists, the expression of the Russian Self inherent in their texts meant that the re-articulation of Russian identity in the official 1999 narrative found fertile soil. In many ways official representations in 1999 projected Russia as a 'prisoner of the Caucasus': not in the classical double sense, but as a pure victim, quite similar to official language during the First Chechen War. However, the view, prevalent in official language during the 1990s, of the Russian state as passive, impotent and weak was replaced in 1999 by articulations resonating with those of the nationalists and communists.

Official representations in 1999 depicted constant threat and attack from the outside as recurrent phenomena in Russian history, while also highlighting Russia's strength and ability to withstand these threats. These representations resemble nationalist and communist accounts of Russian history as a constant struggle to secure its natural hegemonic position as a Eurasian power. The question of guilt is also connected to the idea of Russia under attack. In 1999, the official answer to this question was similar to the position taken by the nationalists. In certain respects, Putin's argument that 'Russia is defending itself: we have been attacked. Therefore we need to throw off all our complexes, also our complex of guilt'[48] echoes Zhirinovsky's stance: 'We have suffered enough. We should make other people suffer.'[49] Zhirinovsky's representation of Russia as 'order' juxtaposed to the criminal and conflict-ridden South also finds a parallel in 1999 claims that Russia was establishing 'order' and 'discipline' in its response to the terrorist threat.

The idea of *unity* is the most striking example of how the 1999 official narrative corresponds with the nationalist and communist position on Russia during the 1990s. Observing the similarities between what he termed the Bolshevist position and the Romantic nationalist position, Neumann noted that both see the links holding 'us' together to be organic, and thus the:

natural and indeed only possible formation and aggregation of the body politic to be harmonious ... the organic metaphor also suggests that any conflict inside the body politic is by its very nature an illness or a disease – an unnatural mode of operation possibly with external causes.[50]

My short re-visit of communist and nationalist texts in the 1990s confirms the strong standing of the organic metaphor. Putin's appeals for unity in the face of the terrorist threat in autumn 1999, postulating harmony between different institutions, the people and the government, spoke directly to this organic metaphor.

The Eurasian position did make certain inroads into official articulations of Russian identity before 1999 – primarily with the introduction of a foreign policy oriented not exclusively toward the West, but also towards the East and South. From the mid-1990s, official discourse started to incorporate language and metaphors of geopolitics from the New Right via the 'democratic statists' who advocated a strong state in combination with a commitment to Western-style democracy. The Near Abroad was represented as crucial to Russia's geopolitical interests and as bound up with great-powerness or national greatness. There was also a retreat from Atlanticism in the sense that a more sceptical view of the USA was articulated.[51] This added a complementary identity for Russia on top of the identity most strongly articulated by the Yeltsin regime, which emphasized Russia struggling to catch up with the West – where it was seen as belonging.

However, it is only with the launching of the Second Chechen War that a confident and positive articulation of Russian identity conquered official language. In the 1999 narrative, the line between 'Self' and 'Other' was clear-cut, and the image of the Russian political unit was one of unity and strength. This was also an official representation of Russia much more in line with that of the political opposition than Yeltsin's ever was. Drawing the lines even further back, we can say that the war was an opportunity to define Russia closer to the Romantic nationalist position. As such, the articulation of Russia implied by the 1999 securitizing narrative marked a first step towards resolving the identity crisis of Russian politics in the 1990s.

Conclusion

This genealogy of 'Chechnya' and 'Russia' in Russian texts has shown that the 1999 official securitizing narrative was not launched into empty discursive space, but resonated with, or refuted, various representations in a mould that was almost two centuries old. Official discourse creates its own content, but also draws on the larger foundation of earlier intellectual and political debates. Several of the basic elements in the new official articulation of Chechnya and Russia already existed somewhere in the bowels of the debate.

As noted in Chapter 4, Putin's 1999 language was foregrounded in media and FSB representations during the interwar period. But it also drew on parts of the classical literary discourse on the Caucasus, blended into historical and more recent accounts on Chechen banditry and criminality, and it drew on positions articulated by the political opposition in the 1990s. In sum, the securitizing move advanced by the Russian leadership during summer 1999 was not alien, but spoke directly to several well-established representations in the Russian discursive terrain. Along with consistent and convincing official argumentation, this deep resonance endowed the 1999 official call for war with a particular appeal to Russian audiences.

While the next chapters investigate the shifting representations of Chechnya and Russia in different audience groups in the course of autumn 1999, the linguistic patterns constructing the Chechen–Russian relation as juxtaposition, repeated over a long time-span, are an important backdrop to these. Such ingrained constructions provide a reservoir on which official calls can draw when addressing the audience. Official language could have played on this reservoir of historical representations during the First Chechen War also, but it failed to do so in a skilful or consistent way. Basically, war is easier to accept when it is waged against an adversary constructed as 'different' and 'dangerous' in many different layers of text over long periods of time and when the call for war is formulated within the boundaries of these identity constructions. As we have seen, the 1999 official call for war was just that. Nevertheless, even when the narrative in a securitizing move speaks to certain well-established representations in the historical discursive terrain, audience acceptance cannot be taken for granted. The narrative can always be changed, appropriated or negated in representations of the putative 'audience' in a broader public debate. This is the topic we turn to next.

Notes

1 H. Ram. 1999. *Prisoners of the Caucasus: Literary myths and media representations of the Chechen conflict*. Berkeley Program in Soviet and Post-Soviet Studies, Working Paper Series. Berkeley, CA: University of California, 3.
2 The poem 'Kavkazskiy Plennik' (Prisoner of the Caucasus) was written by Aleksandr Pushkin in 1822; both the title and theme have been recurrent in Russian literature ever since, recently in Sergey Bodrov's film *Prisoner of the Mountains* (Orion Pictures Corporation, 1996).
3 Ram 1999, 11.
4 S. Croft. 2012. *Securitizing Islam: Identity and the search for security*. Cambridge: Cambridge University Press, 90.
5 O. Bullough. 2010. *Let our Fame be Great: Journeys among the defiant people of the Caucasus*. London: Penguin, 261, 313.
6 Ibid., 154, 194, 195, 204, 209, 217.
7 Ibid., 154.
8 Vera Tolz (1998. Forging the nation: National identity and national building in post-Communist Russia. *Europe–Asia Studies*, 50(6): 993–1022, 1003) writes that these representations were not confined to Samizdat, but were also to some extent reflected in such official journals as *Molodaya Gvardiya* and *Nash Sovremennik*.
9 Tolz (1998, 1004) mentions the National Republican Party of Russia and Russian National Unity as the two main racist groups in post-communist Russia.
10 Ibid.
11 J. Russell. 2002. Mujahedeen, mafia, madmen ... Russian perceptions of Chechens during the wars in Chechnya, 1994–1996 and 1999 to date. *Journal of Communist Studies and Transition Politics*, 18(1): 73–96.

12 Ram 1999, 15–18.
13 B.N. Yeltsin, Annual Speech to Deputies of the Federal Assembly (16 February, 1995) listed in *Russia and Eurasia Documents Annual 1995*, available at www.ai press.com/REDA.contents.95.1.html, and accessed 29 February 2016.
14 In Chechen discourse, in fact, it is largely positive. The wolf is the national symbol. It features in the national anthem and under a full moon on the flag of the republic of Ichkeriya/independent Chechnya.
15 J. Russell. 2005. Terrorists, bandits, spooks and thieves: Russian demonisation of the Chechens prior to and since 9/11. *Third World Quarterly*, 26(1): 101–116, 106.
16 Ram 1999, 16.
17 Ibid., 18.
18 A. Kassianova. 2001. Russia: Still open to the West? Evolution of the state identity in the foreign policy and security discourse. *Europe–Asia Studies*, 53(6): 821–839.
19 E. Mikiewicz. 1997. *Changing Channels: Television and the struggle for power in Russia.* New York: Oxford University Press; E. Mikiewicz. 2008. *Television, Power, and the Public in Russia.* Cambridge: Cambridge University Press.
20 Ram 1999, 22.
21 Ibid., 21.
22 Ibid., 2. John Russell's study (2002, 84) drew similar conclusions on media representations of Chechen warlords during the First Chechen War:

> The part played by Basaev and Gelaev in the final rout of the federal forces in Grozny in August 1996 served to heighten their prestige as national heroes in the eyes of the Chechen people and as daring 'Robin Hood' revolutionaries by broad sections of the Russian media.

Artistic films produced in this period such as Sergey Bodrov's *Prisoner of the Mountains* also portrayed the Chechen as a Noble Savage and not as a wild and violent enemy.
23 The original team consisted of Kovalev, State Duma Deputies Valery Borshchev, Mikhail Molostvov, and Leonid Petrovsky, and expert at the Memorial Society's Human Rights Centre Oleg Orlov. They were later joined by Deputies Yuly Rybakov and Aleksandr Osovtsov.
24 The reports are cited in E. Kline. 1995. *Chechen History.* [online] Available at www.newsbee.net/moscow/chhistory.html and accessed 26 July 2014.
25 Ibid.
26 Translated and quoted in Ram 1999, 3.
27 There are two wings within this position, the xenophobic and the spiritual, the latter of which, according to Neumann, was almost crowded out of the debate in the 1990s (I.B. Neumann. 1996. *Russia and the Idea of Europe.* London: Routledge, chapter 8). The key element in the Romantic nationalist position is:

> the organic nation, understood as a living being where each part is dependent on the others, and where no basic conflict of interest can therefore exist. The state is seen as the head of the organic nation, embodying its will, defining its interests and defending it against harmful internal microbes and external onslaughts. The wellbeing and good fortune of nation and state are guaranteed by God or a functional equivalent thereof – for example, the course of history.
>
> (Neumann 1996, 196)

28 Tolz 1998, 1012.
29 Ibid., 996.
30 L. March. 2001. For victory? The crises and dilemmas of the Communist Party of the Russian Federation. *Europe–Asia Studies*, 53(2): 263–290: 270. On the sources that contributed to Zyuganov's texts, March writes:

The basic contours of this ideology are well known. In both form and language it is derived from nineteenth-century Russian conservative thought, the anti-fascist fronts used from 1942 onwards by the Comintern, the national communist ideology of the Great Patriotic War.

31 According to the 'Eurasian Idea', Russia should follow its own distinctive societal and geopolitical path separately from Europe and the West. In its new, early 1990s version, Russia is seen as the leading Eurasian state with a special role within the post-Soviet space.

32 G. Smith. 1999. The masks of Proteus: Russia, geopolitical shift and the new Eurasianism. *Transactions of the Institute of British Geographers, New Series*, 24(4), 481–494: 486.

33 Drawing this conclusion from Zyuganov's texts, Andrey P. Tsygankov places him amongst what he calls the 'aggressive realists' in Russia (A.P. Tsygankov. 1997. From international institutionalism to revolutionary expansionism: The foreign policy discourse of contemporary Russia. *Mershon International Studies Review*, 41(2): 247–268, 256).

34 A. Ingram. 1999. A nation split into fragments: The Congress of Russian Communities and Russian nationalist ideology. *Europe–Asia Studies*, 51(4): 687–704, 700.

35 From Zyuganov's book *Rossiya i Sovremennyy Mir* and cited in Graham Smith (1999, 486).

36 G.A. Zyuganov. 1992. *Drama Vlasti*. Moscow: 184.

37 J.B. Urban and V.D. Solovei. 1997. *Russia's Communists at the Crossroads*. Boulder, CO: Westview Press, 100.

38 Both Aleksandr Prokhanov and Aleksandr Dugin were central thinkers within Russia's New Right. Their writings were primarily published in journals such as *Zavtra* and *Elementy: Evraziyskoe Obozrenie*.

39 Smith 1999, 485.

40 Ibid.

41 Prokhanov, quoted in Neumann 1996, 186.

42 A nationalist organization led by Dmitry Rogozin which came to prominence during the Duma elections in 1995, but failed to cross the 5 per cent threshold for federal list representation. Yury Skokov and Aleksandr Lebed were also recruited to KRO.

43   We were a united nation and we shall return to national unity. Only having overcome the division of the russkaya natsiya is it possible to restore civil dignity to millions of people, to revive Russia and save her priceless culture from annihilation.

                            (Manifesto of KRO, cited in Ingram 1999, 690)

44 Cited in Ingram 1999, 701.

45 Ibid.

46 Smith 1999, 484–485.

47 LDPR came out as the victor in the 1993 Duma elections with 22.92 per cent of the vote, whereas the CPRF emerged as the victor in 1995 with 22.30 per cent.

48 'Kreml' izbavlyaetsya ot kompleksa viny', *NeGa*, 9 September 1999.

49 Cited in Ingram 1999, 701.

50 Neumann 1996, 174.

51 In what Smith (1999) refers to as 'Official Eurasianism'.

7 Political elite representations of Chechnya and Russia

In this book, the process combining securitizing moves and audience acceptance is theorized as an intersubjective process of legitimation. As noted in the theory chapter, the production of a consenting audience, which leads to acceptance of emergency measures beyond rules that would otherwise have to be obeyed, is seen as a *joint act* in which securitizing actors and audience participate (Chapter 2). Once the securitizing move has been launched, the reception of the securitizing narrative is shaped by the discursive terrain already existing among the audience, while there is also room for change in the discursive terrain and appropriation of the narrative. 'Audience acceptance' does not happen at one point in time or one moment: it is an ongoing process of legitimation through which the representation of something as an existential threat acquires a hegemonic position at the expense of other, less threatening, representations. This 'happens' when the description of the threat as 'existential' and of 'the point of no return' and 'way out' indicated in a securitizing move has acquired sufficient resonance in representations of the audience to enable emergency action to be undertaken *legitimately*.

Obviously, an empirical study cannot fully capture the dynamic social processes suggested in this explication. However, we can get an idea of certain aspects of intersubjective dynamics (change and appropriation of the narrative) by studying *changes* in audience representations over a given timespan: here, September through to December 1999. In the next three chapters, I focus on revealing how the intersubjective process unfolded by investigating similarities, differences and changes in representations in and across the texts of various audience groups and comparing these to the official narrative extracted in Chapter 5. This makes it possible to establish how far the process of producing a consenting audience evolved during autumn 1999 and how it happened.

This chapter presents the texts of members of the Russian political elite who were not in government, but who held or campaigned for seats in the Federal Assembly of Russia.[1] As shown in Chapter 6, the new 1999 official representations of Chechnya and Russia fitted certain positions in the Russian discursive terrain fairly well. In particular Putin's imagery of Russia resonated with dominant representations among the CPRF and the New Right. Moreover, representations of Chechnya as a dangerous Other were nothing new in Russian discourse. This fertile discursive terrain certainly worked towards 'acceptance' of the official narrative by the Russian political elite during autumn 1999. On the other hand, the analysis of elite discourse below shows that the process that led up to agreement on the gravity of the Chechen threat and the necessity of a new war was indeed an intersubjective one. Putin's narrative was not only replicated, but also reformulated and accentuated in the representations offered by members of the Russian Federal Assembly that autumn.

In the following, I uncover this process by focusing on the extent to which representations of the Chechen 'threat' (including representations of Maskhadov), 'the point of no return' and 'the way out', as well as representations of 'Russia' given by Federation Council and State Duma members during autumn 1999, overlap with those in the 1999 official narrative. I first trace how the alternative position on Chechnya, identified as the 'discourse of reconciliation' in the interwar period, all but disappeared from the language of members of the Russian Federal Assembly during autumn 1999. Then I move on to what emerged as the dominant position in political elite discourse: how it matched and underscored official claims about the new relation between Chechnya and Russia and the most appropriate 'way out', but also how it differs from those.

The focus in this chapter is on linguistic narratives. However, the abstract situation of a policy being established as legitimate also has implications for *formal acts* undertaken as part of the 'emergency measures'. Once the legitimacy of a policy has been established, this may lead to its being legally and formally authorized. Thus seen, audience acceptance entails an emerging overlap in representations, an overlap that also finds its expression in concrete formal acts such as passing a law or agreeing to a change of policy. The former will be explored in this chapter; the latter, in Chapter 10.

The waning of an alternative position on Chechnya

During autumn 1999, representations of threat, blame and the 'way out' among the Russian political elite were not identical to official representations. Initially, in the emerging debate on terrorism and Chechnya that autumn, there were alternative positions to that offered in official language. Where the official narrative underscored the danger of the Chechen threat by emphasizing it as inhuman, barbaric and violent, this alternative position constructed Chechnya, or at least the Maskhadov regime, as human and reasonable, with Russia as the guilty party. This position was voiced by Aleksandr Lebed, broker of the Khasavyurt Accord and now governor of Krasnoyarsk, and, albeit much less vocally, by the head of the Our Fatherland Party Yevgeny Primakov.[2] Lebed indicated that the Russian powerholders were directly responsible for the terror, and portrayed the Chechen warlords as decent and human.[3] Former head of the Supreme Soviet Ruslan Khasbulatov also intimated that Russian powerholders needed a war in the Caucasus to demonstrate strength prior to the elections.[4]

The representations of former CIS Secretary Boris Berezovsky also belong to this position, at least in terms of placing the blame on Russia. Although he said that Chechnya was the source of the explosions and indicated that Chechnya was closely connected to international terrorism of the fundamentalist Islamic strand, he accused the Russian authorities of contributing to this development by neglecting Chechnya in the interwar period.[5] Similarly, although the President of Bashkortostan Murtaza Rakhimov proposed that the terrorists should be isolated and that the harshest measures possible be undertaken against them, he also claimed that the Russian government was responsible for the situation because they used guns against their own population in the North Caucasus. His proposal for a 'way out' was to stop military action in Dagestan and Chechnya and sit down at the negotiating table.[6]

Within this alternative position, 'Maskhadov' was never detached from his identity as the legitimately elected president of Chechnya. On the whole he was given a very different identity from that indicated in Russian official rhetoric. Even some of Zyuganov's statements must be placed within this position on this particular point. On 28 October, Zyuganov indicated that the President's policy in Chechnya was 'criminal' and would lead to the 'final collapse of the Russian Federation' because 'the present Russian regime had financed Dudayev and did nothing to negotiate cooperation with President Maskhadov'.[7] Former Prime Minister Sergey Stepashin commented on Putin's controversial statement on 1 October that the Chechen parliament of 1996 was the only legitimate organ of power in Chechnya:

I would not burn all our bridges with Maskhadov here. We have put ourselves in a delicate situation. The agreement was signed by Yeltsin and Maskhadov. We acknowledged him as a legitimate president. There shouldn't be any double standards! You should leave yourself a small loop hole at the very least! You can't corner people and at the same time try to reach an agreement with them.[8]

Yabloko leader Grigory Yavlinsky offered similar representations of Maskhadov:

Russia has the President whom it elected in 1996. Maskhadov was elected in the same way. In this sense he is also a legitimate president. Also, Maskhadov has one advantage over everybody else in Chechnya – he is not connected with Moscow's political criminal circles.[9]

Within this alternative position, then, 'Maskhadov' was represented more as part of the Russian Self, than as part of a threatening Other.

Although there clearly were variations within this alternative position, it deviated from the official position by underscoring the legality of Maskhadov and offering a sharper distinction between 'terrorists' and 'Chechens'. For example, the statements of Primakov emphasized the 'Chechens' as reasonable, human and close to Russia. He indicated that the Chechens themselves would eventually fight extremism:

There will be more and more people who regard the fight against extreme elements as their duty for survival and welfare.... Executing wide land operations, which would develop into a full-scale war, by contrast, would impede the creation and strengthening of the healthy elements in Chechnya itself.[10]

The appropriate 'way out'/policies accompanying these more benign representations of 'Chechnya' were those of communication and negotiation, with a corresponding rejection of an all-out war. Yevgeny Primakov argued against a full-scale war: 'I am categorically against this.... This cannot lead to any positive outcome. Instead there will be a lot of casualties, both among the civilian population and among our soldiers.'[11] Grigory Yavlinsky's 6 October proposal of an official meeting with Aslan Maskhadov parallel with the armed operations was a logical fit with the identity construction of 'Maskhadov' within the alternative position.[12] Finally, on the question of who was to blame and the articulation of the Russian Self, this position indicated Russia as guilty – whereas official rhetoric had presented Russia as strong and innocent.

The alternative position clearly builds on the interwar 'discourse of reconciliation' – indeed, it was articulated by many of the same people. However, the quotes above show how the discourse representing Chechnya as an existential terrorist threat was making inroads into this position. The Rakhimov quote in particular contains an uneasy combination of both positions: he names Russia as the culprit and calls for political solutions, but also endorses the terrorist talk and the accompanying promotion of violent measures. The statements of President Aleksandr Dzasokhov of North Ossetia seem to try to accommodate both the official and the alternative positions:

the events in Dagestan again shed light on the huge danger our state is facing ... the necessity of undertaking radical measures against armed extremism, and eradicate the root causes behind the huge armed hotbed which had emerged in Dagestan ... a political solution to the Chechen problem is a first priority.[13]

As early as mid-September, then, representations of Chechnya as an existential terrorist threat necessitating violent response were incorporated into and weakening the alternative position.

This process is best illustrated by studying the changing representations of 'Chechnya' offered by of the head of the Yabloko faction Grigory Yavlinsky. While his initial statements could be placed within the alternative position, his language quickly moved to accommodate the official securitizing narrative. As early as the end of September, Yavlinsky's language incorporated both positions: 'We should ruthlessly eliminate bandits and their groups and be extremely careful with civilians, as we are with compatriots who are in danger. Only in this way can we finally achieve positive results in the Northern Caucasus.'[14]

When Yavlinsky, in a declaration to the Duma on 9 November, proposed that negotiations should be conducted with 'Maskhadov as the legitimately elected President of the Chechen Republic' (alternative position), he did this by first expressing complete support for the 'way out' suggested in the official narrative and already implemented against Chechnya through continuous bombing, a full ground offensive, *zachistki*, etc.[15] He noted: 'the Russian army has completed its task in the Northern Caucasus, creating for the first time in the past five years a convincing prerequisite for a political settlement of the problems there'. Similarly, his proposal of declaring a state of emergency in Stavropol, Dagestan and other territories bordering Chechnya 'to ensure the required minimum legal basis for the actions of the military forces of the Russian Federation' were justified not with reference to protecting the rights of the civilian population, but 'to protect the security of Russian citizens and secure strategic state interests'. The very harsh terms for negotiations with Maskhadov indicated in Yavlinsky's declaration were quite similar to those stated by Putin back in September, and the wording linked 'Maskhadov' to terms such as 'hostages', non-existence of 'a state governed by civil law', 'kidnapping', 'slave trade', 'terrorists' and 'terrorism'. Further, the declaration stated that, if Maskhadov could not manage to rid Chechnya of all these problems, 'a 30-day deadline should be granted to enable all refugees to leave the Chechen republic. Then the aforementioned tasks will be solved by the federal forces independently'.[16] Also the articulation of Russian identity in Yavlinsky's language seemed to resonate with that in the official narrative:

> I would prefer to have a better trained, better paid and better equipped army, as Russia is a country that can either be strong and powerful or cannot exist, and it will be torn into pieces ... there is no other way out. Look at our borders.'[17]

Thus, the alternative position was all but subdued by the official one in Yavlinsky's language. The fact that even the wording of his 9 November declaration spurred one of Russia's best-known liberals Anatoly Chubays to brand Yavlinsky a 'traitor' because 'implementation of Yavlinsky's plan would virtually not only stab the Russian army in the back, but also help Maskhadov evacuate the terrorists beyond the borders of Chechnya and hide them from justice' is an indication of how normalized the representation of 'Chechnya' and 'Maskhadov' as an existential terrorist threat had become and of how accepted the emergency actions undertaken by the Russian government were among the Russian political elite. Equally telling was the defence of Yavlinsky by his fellow Duma representative Alexey Melnikov, who stressed that Yavlinsky 'supported and would support the actions of the Russian army in Chechnya ... Yavlinsky's plan did not envisage any removal of the blockade on Chechnya or harbouring of international terrorists from justice'. According to Melnikov, 'the plan seeks to remove the terrorists with minimum losses for Russian troops and ensure a political settlement of the situation from a position of force'.[18]

Linking back to the theory-based explication of securitization theory (Chapter 2), the process described above shows how audience representations can reject or reformulate – but also appropriate – the securitizing narrative once it has been launched.

The waning of the alternative position on Chechnya among the Russian political elite during autumn 1999 was visible not only in the changing pattern of speech among a few liberal politicians. It could also be read from the pre-election campaign. Among all the parties and politicians that could have voiced criticisms of the war as a means of mobilization, the alternative discourse on Chechnya was virtually non-existent.[19] Instead, most statements by the Russian political elite that autumn served to reinforce the official securitization of the Chechen threat. And to that we now turn.

... and reinforcement of the official position on Chechnya

While an alternative position on Chechnya was voiced among the Russian political elite in the beginning of autumn 1999, representations in line with the official language were much more widespread. Predictably, the official 1999 narrative for war was echoed in statements given by the well-known securitizing voice from the interwar period, that of Anatoly Kulikov, who was now campaigning for a seat in the Duma. During a press conference on 10 September he stated:

> there should be no negotiations with Basayev and Khattab. To talk to bandits is useless ... we need to destroy the fighters fully and without any losses on our side.... I am categorically against the independence of Chechnya.... It is not a secret that they receive money from international terrorist organizations.[20]

With the labelling of the threat as 'bandit' and at the same time invoking the distant enemy by referring to 'international terrorist organizations', Kulikov's description of the threat resembled that of the official narrative and also suggested a policy of destruction and non-negotiation. However, the securitizing narrative indicated by the Russian leadership during summer and autumn 1999 was echoed by a much wider circle of people than traditional hawks in the Russian political elite. Kulikov's words, which had seemed so out of touch during the interwar period, became mainstream in Russian elite discourse during autumn 1999.

Sergey Stepashin, for example, who had offered representations of Chechnya quite different from those in Putin's language only a few months earlier, in an interview on 18 September offered a representation of Chechnya and Russia very similar to those of the official narrative:

> Finally, Russia has to learn to count and Chechnya to pay its dues. They have something to pay. Stealing oil, dollars that are used to buy weapons....

When they talk about 180 billion that Russia should pay for the war ... they should pay us for the war! We didn't start the war and anyway 98 per cent of the infrastructure of Chechnya was built by the Soviet Union.... It is necessary to know these bandits, they take the money. They will take a lot of it if you offer it, but they will act as they want to. To them we are 'dogs': it is possible to kill us, cut off fingers, heads.

As to the 'way out': 'I am for the harsh measures that are used today against the band formations.'[21] In this text, Chechnya is represented as an unreliable villain; guilt is placed squarely on the Chechen side, while Russia is represented as innocent – even, in the form of the Soviet Union, as the sole source of order and civilization in Chechnya.

In general, widespread agreement on Chechnya as an *existential terrorist threat* and on Russia as standing at the *point of no return* was developing in elite discourse. Statements repeatedly represented Chechnya as 'terrorist' 'bandit' or a 'hotbed of armed extremism'. Chair of the Committee on Defence in the Duma Roman Popkovich described Chechnya as a 'criminal state, a centre of terrorism not only in the North Caucasus but in the whole of the Middle East'.[22] The gravity of the threat was underscored by referring to the situation as 'total terror'[23] or more frequently as a 'war'[24] and the drawing of parallels between the Second World War and the present situation.[25] Even if Communist Party representatives continued to securitize the Yeltsin regime, Chechnya was now represented as the most violent and immediate threat within this regime. In the words of Communist Party leader Zyuganov, Chechnya was a:

more terrible manifestation of the illness of the whole state and social organism ... Chechnya is not the primary source of infection but its most violent symptom.... The terrorist Chechen regime is an undivided part of the Yeltsin criminal oligarchic regime, which reigns Russia.[26]

Despite lingering criticism of the Yeltsin regime, the similarities between official statements on the terrorist threat and those of most of the political opposition were striking. Just as in the official narrative, the dominant elite discourse now transmitted the impression that the terrorist danger was about to engulf Russia entirely. As early as the incursion into Dagestan, Vladimir Zorin, Chair of the Duma committee on nationalities and member of the Nash Dom–Rossiya Party, had said: 'If we do not stop this conflict now, then the whole country might be dragged into it. The whole society must understand the danger of terrorism. For Russia it is problem Number 1.'[27] And similarly the Chairman of the Federal Assembly Federation Council Yegor Stroyev (CPRF): 'terrorism has become a daily reality and Moscow is not secured against it, nor are any regions of Russia'.[28] In terms not only of space but also of time, the threat was constructed as overwhelming. According to Duma deputy Nikolay Ryzhkov: 'Russia will have to live with the problem of terrorism for many, maybe even tens of, years to come. We have to be psychologically ready as the threat will not go away right now and there is no simple solution.'[29]

The 'Chechens', their president 'Maskhadov' and the 'distant enemy'

While a distinction was usually made between Chechnya/terrorism as a security threat on the one hand and Chechens/North Caucasians on the other, this was not always the case, with 'Chechen' sometimes occurring in the same sentence as 'terrorism', or even more directly, as when the head of the Duma Security Committee Viktor Ilyukhin (CPRF) stated that the responsibility for the terrorist acts must be put on 'representatives of Caucasian nationality', of which there were more than a million living in Moscow because of 'neglect by the government (power)'.[30] Such talk contributed to constructing the Caucasian people as a security threat in themselves. On the whole, the securitization of 'Chechnya' as an existential threat easily slipped over into giving 'Chechen' the same meaning. Often, the co-existence within the same text of representations that served to dehumanize and securitize 'Chechnya' by linking it to terms such as 'killing of civilians', 'taking of hostages', 'terrorist' and 'criminal filth' and representations that sought to de-securitize 'Chechen' with phrases such as 'the habits of the mountain dwellers must be respected ... the Chechen people merit respect' resulted in a contradictory construction of 'Chechen'.[31] Moreover, with the enormous amount of securitizing talk, the smaller story of the 'good Chechens' that could make up a part of the Russian 'Self' somehow seemed to get lost.

What then of the Ichkerian President Maskhadov? If we study 'Maskhadov' as an 'event within the event', as was done in the analysis of official language in Chapter 5, 'Maskhadov' in the language of most of the political elite was gradually moved from the position of a legitimate and trustworthy partner, to that of an unreliable and weak individual, potentially an accomplice of the terrorists. Several statements immediately dovetailed with Putin's initial framing of Maskhadov as consenting to terrorism. State Duma Defence Committee Chairman Roman Popkovich declared that, if Maskhadov was incapable of disbanding the guerrillas, he should step down and make way for a new government. Ruslan Aushev proposed that Moscow should co-opt those forces in Chechnya that also 'seek to fight terrorism', although he did not mention Maskhadov by name.[32] Gennady Zyuganov, who even said that Moscow should have supported Maskhadov's government to a greater degree than it had done, now cast some doubts on Maskhadov's credentials, criticizing him for not having apologized to the Dagestanis for the incursion of Chechen fighters into the republic in August.[33]

Quite controversial was Putin's statement on 1 October that the Chechen Parliament of 1996 was the only legitimate organ of power in Chechnya – implying that Aslan Maskhadov was not the legitimately elected President of Chechnya. At the time, no-one in the Russian political elite had yet expressed doubts as to the legitimacy of Maskhadov as Chechnya's president. And, as noted, this new representation of Maskhadov was not immediately accepted by everyone (see alternative position above).

However, the balance was tipping in favour of downgrading Maskhadov. Moscow Mayor Luzhkov declared: 'not one of the current organs of power in Chechnya can be considered legitimate ... Maskhadov does not recognize the Russian Federation and the Russian Federation does not recognize him.' And Sergey Sobyanin, Chair of the Committee on Constitutional Laws in the Federation Council, contended that according to Russian law 'the current Chechen president is not the legitimately elected president of the republic, because he was elected according to their Chechen laws and not the Russian laws'.[34] Here we see that a clear boundary was being drawn, separating Maskhadov both from 'Russia' and from the orbit of legality. The fact that not one of the key politicians (former prime ministers and heads of key Duma factions) present at the meeting with Putin on 5 October, when it was decided whom to ask to serve as the new 'general governor' of Chechnya, defended Maskhadov as the president of Chechnya indicates how dominant this new representation had become.[35] The idea of a 'political solution' to the Chechen problem was present in official discourse as well as among the political elite throughout that autumn. 'Maskhadov', judging by the changing representations, no longer looked like someone who could take part in such a process.[36]

While political elite representations served to amplify the official identification of Maskhadov as an unreliable partner, they did not coincide with other core parts of the official narrative. The official narrative had indicated that 'Osama bin Laden' or 'enemy circles in Muslim countries' stood behind the Chechen threat as a distant enemy. This was not a widespread representation in the Russian political elite as such at the time: if a 'distant enemy' was suggested, it was rather the USA. In an extensive opinion piece by Duma deputy Nikolay Ryzhkov, for example, the USA is represented as an expansive and aggressive power; Ryzhkov indicates that:

the USA is trying to exploit Islam's energy for its geopolitical goals. Formally against fundamental Islam, the Americans, in essence, are sending extremists against their rivals – in particular against Russia and increasingly against Europe, creating an 'iron curtain' of instability in Southern Eurasia.[37]

Anti-American/anti-Western discourse remained a consistent feature of Communist Party discourse, often intertwined into representations of Chechnya as well.

The 'way out'

Whatever mismatch there might have been between official discourse and elite discourse on the specific features of the threat, there was agreement not only that the threat was existential and that Russia was standing at the point of no return, but also, as it turned out, on the *possible way out*, on the appropriate means to undertake in order to fight off the threat. The overwhelming majority of Duma representatives were reported to have supported the 'strong hand' approach of Vladimir Putin following his address to the Duma on 14 September and the Federation Council on 17 September.[38] Support for the emergency measures proposed in the official narrative was evident amongst several senators as well. Indeed, the press reported that there was an atmosphere of consensus on how to deal with the security challenge during the session in the Federation Council, with Senators describing the government's handling of events in Dagestan and Moscow as 'sensible'.[39] This crude indication of elite 'acceptance' of the emergency measures proposed by the Russian leadership in autumn 1999 for fighting the Chechen threat is confirmed if we examine the language employed by the political elite. St Petersburg governor Vladimir Yakovlev said that Putin's presentation was 'hard, but right'; the Ingush President Ruslan Aushev, known for his critical views on Russian policies toward Chechnya, was reported to have expressed support for 'a struggle without compromises against terrorism, extremism and banditry in Russia'.[40] And Vladimir Zorin stated:

I have always been an advocate of political means for solving problems, but this time I support the determined actions of the Russian leadership as the only possible ones. Terrorism merits one fate – liquidation. In this respect there can be no other options.[41]

If anything, most statements by Federal Assembly representatives seemed to suggest measures even further beyond 'rules that otherwise have to be obeyed' than those indicated in the official narrative. For example, State Duma Speaker Gennady Seleznev (CPRF) said that Russian troops had the right to annihilate guerrillas on Chechen territory; further, that Moscow should ignore European pressure to abolish capital punishment and sentence the guerrilla leaders to death.[42] Others argued, long before the Russian leadership launched such an idea, that air strikes were insufficient and that ground troops should be sent into Chechnya, suggesting an all-out war against Chechnya.[43] Thus, by the end of September, consensus had emerged on the controversial (due to the First Chechen War) question of adding a ground offensive to the bombings of Chechen territory. Putin's 'I never said there would not be a ground offensive' was matched by the words of Head of the Federation Council Yegor Stroyev: 'the Terek river [running through Chechen territory] is a good barrier', and Head of the Duma defence committee Roman Popkovich's statement, 'we need to get under our control some operative territory from which to fend off counterattacks from the fighters'.[44]

The logical flipside of proposing violent measures for dealing with Chechnya was the rejection of such policies as negotiation and cooperation. According to the Federation Council, the Khasavyurt Accord – the very symbol of peace and reconciliation with Chechnya from the interwar period – 'had caused huge damage to the security of the Russian Federation'.[45] The strong criticism of the

Khasavyurt Accord and the 1997 peace agreement from across the political spectrum made clear the irrelevance of the interwar de-securitization discourse.[46] With the rejection of negotiation as a 'way out' came the denunciation of those who advocated such policies, as when Viktor Chernomyrdin stated:

> I categorically condemn those of Russia's internal forces who conform to anti-Russian Western circles, dramatize the hysteria around the 'humanitarian catastrophe', and call for a halt to military operations and starting the negotiations.... Negotiations are not carried out with bandits. Bandits are killed, for those who want to live and work normally.[47]

Or, in the words of Governor of Saratov Dmitry Ayatskov:

> Our problem is that we behave like Tolstovians; we excuse bandits, drug barons, traitors, we give amnesty to those who steal from and degrade our great nation and tolerate deceitful Judases, talkative idle doers, at any time ready to sell themselves and their country for thirty silver coins. But it is necessary to destroy physically the first [group], send to prison the second and expel the third, just as they do with their enemies, traitors and criminals in countries with self-respect.[48]

On the whole, agreement emerged in autumn 1999 between the Russian leadership and the broader political elite in the Russian Federal Assembly on the severity of the Chechen threat and on the need to adopt force to counter the threat, leaving behind the policies of peace and negotiation. A further new similarity between official and elite discourse was the call for Russian *unity* as part of the 'way out', as a prerequisite for withstanding the terrorist threat. Just as Vladimir Putin had done, head of the Federal Assembly Yegor Stroyev argued that there was a need to unite the regional and federal levels to fend off the terrorist threat.[49] Communist Party leader Gennady Zyuganov immediately proposed several measures aimed at 'unifying' power in Russia. His proposals were to 'strengthen all security agencies', 'stop the reshuffling of cadres in government' and 'insist on holding joint sessions with both chambers of the Federal Assembly'.[50] The strong emphasis in Zyuganov's language on *unity* as a prime value both in the organization of state power and of territory thus both preceded the official discourse of autumn 1999 (as shown in Chapter 6) and reinforced it during the first months of the war. Statements such as 'questions of national security of Russia and pursuing its state unity and sovereign rights in the whole territory of the country have incontestable priority in comparison to regional problems'[51] indicated acceptance of the official position on Chechnya. Repeated over time, they also served to build legitimacy around this position.

This common call concerned not only unity across the regional/federal divide, but also unity across the divide between regime and society.[52] According to Duma deputy Nikolay Ryzhkov, 'the whole world experience about the struggle against terrorism is based on the mutual actions of the power and the population

... we should immediately develop the national propaganda of methods of struggle against terrorism'.[53] And Dmitry Ayatskov, Governor of Saratov, wrote:

I would like our constitution to correspond with the status of a law-based great power and that Russia could stay great and undivided, and that Russians could be proud of their country. We will survive and overcome all problems if we understand: the question is not who is more important or influential right now nor who has the right political affiliation, but how we can save Russia. The risks are too big right now that in the next century Russia in its current shape will cease to exist. Not one powerful state in history has survived when the central power is weak and people and army are left to live in economic, political and legal chaos.[54]

Thus, the official call for 'unity' as a means of securing Russia against the terrorist threat was reinforced in elite representations. Nor is this surprising, given the prevalence of the unity theme in historical Russian discourse. More important here is that this agreement on the acute need to unite in the face of the terrorist threat served to build the power of the Putin regime in the longer term.

Although some elite statements continued to depict the Russian government as a culprit,[55] the *representation of Russia* as the innocent party to the conflict was becoming fairly widespread, particularly in texts revisiting the interwar period in Chechnya and in texts on the Khasavyurt Accord. Chechnya was depicted as having broken all its promises and Russia as having fulfilled them. Similar to Stepashin's reasoning cited above, Zorin, for example, declared: 'using force is justified because the current Chechen authorities practically repudiated the Khasavyurt Accord. Grozny blamed and blames Moscow for not fulfilling obligations of economic aid and re-building the republic. But it is a myth!' Here the construction of Russia as trustworthy and innocent is underlined by juxtaposing the Russian against the Chechen side, which has 'not confiscated any weapons, or liquidated any criminal gang ... all the time the taking of hostages, killing of civilians and terrorist acts have continued'.[56] On the basis of such identity constructions and the moral juxtaposition of Russia against Chechnya, violent retribution seemed legitimate and logical: 'We have a total constitutional and moral right to create a chain of military and economic blockades and suffocate the fighters. Then there will be hope that a normal life in Chechnya can be built.'[57]

The view of Russia as being morally right is often linked to the idea of Russia as bringing *order and reason* into Chechen chaos. In the words of Vladimir Zorin, 'we are obliged to destroy terrorists, to cleanse the territory of Chechnya from criminal filth (*skverna*) ... the authorities have strongly decided to set up order in our common home'.[58] Similarly, Viktor Chernomyrdin described 'the Chechen republic' as 'part of Russia. Unfortunately, today it is very sick. But Russia has enough reason, force and resources to raise the sick to its feet. We are in a position to put our house in order.'[59] Here also, there was a good new fit between the official discourse on Chechnya and Russia and that of the political elite in the Russian Federal Assembly. This served to reinforce official rhetoric and signified acceptance by the audience across yet another dimension of the securitizing narrative.

As we shall see in Chapters 10, 11 and 12, Russia's innocence and righteousness were also widely insisted on by the Russian political elite as the military operation proceeded and the enormous human costs became evident. Take Vladimir Ryzhkov's statement on Radio Svoboda on 23 November:

> I have never agreed with Russia perpetrating aggression, Russia perpetrating humanitarian terror or such against the civilian population. It is not right. Russia is actually now taking all possible steps in order to get the civilians (*mirnye grazhdane*) out of there.

He repeated this view of Russia in December 1999, but this time juxtaposed it to the West's cruel behaviour in Belgrade and stated that 'Russia, which took in hundreds of thousands of Chechen refugees, hundreds of thousands of Russian refugees from Chechnya, Russia now does everything to restore peaceful, quiet life there.'[60] This articulation of Russia as strong, fair and innocent not only duplicated the new official discourse on Russia, but also stood in stark contrast to the official articulation of Russia during the First Chechen War and indeed in the entire Yeltsin period.

Conclusion

We have seen the impressive degree of overlap between representations of Russia and Chechnya in the texts of the Russian political elite and those of the Russian leadership during autumn 1999. There was widespread agreement on the nature and the gravity of the threat, with several similar terms being used. As to 'Maskhadov' as an 'event within the event', his status as a legitimate figure lingered on in Russian political elite representations, but here also we find no real mismatch with official representations of him.

Although political elite representations did not construct the Chechen threat as part of the international terrorist threat, indicating the West as a 'distant enemy' instead, most of these texts indicated the need to use tough and violent measures against Chechnya as the only possible 'way out'. Some even seemed to indicate the need for more radical measures than those proposed in the official narrative. Finally, the new official articulation of Russian identity, projecting Russia as strong, innocent and capable of establishing order, was repeated during autumn 1999 in the statements voiced by the Russian political elite.

On the whole, we may say that even as information on the heavy human costs of the military campaign was starting to trickle through the increasing barrier of information control, statements among Federal Assembly representatives 'hardened' in the sense that recourse to tough emergency measures beyond 'rules that otherwise have to be obeyed' became accepted, with reference to the unprecedented gravity of the threat and the righteousness of Russia.

And how did this acceptance come about? The review above has made clear the intersubjective nature of the process. Representations prevalent in the language of the national patriotic bloc during the 1990s had been merged into the 1999 official narrative together with more ingrained historical representations of Chechnya. No surprise then that the official narrative had some initial appeal to members of the Federal Assembly. Then again, the confirmation of this official narrative in political elite representations during autumn 1999 was not merely an echo: it was a re-articulation of the official narrative, inserting certain new aspects as to the construction of the threat. This says something about how the new consensus on Chechnya was produced: not so much by command as by common discursive efforts. The net effect of political elite representations on Chechnya and Russia that autumn was to add a further layer to the construction of Chechnya as an existential terrorist threat against Russia. Even the marginal 'discourse of reconciliation' was re-articulated in political elite language in such a way that it helped to confirm the official narrative, rather than contradicting it.

Thus, we can conclude that the language of Russian politicians who held or were campaigning for seats in the Federal Assembly not only indicated that the new war was a legitimate undertaking in the eyes of this crucial audience: it also contributed greatly to constituting Chechnya as an existential terrorist threat and served to substantiate and underscore the official securitizing claim. The statements of the Russian political elite on Chechnya that autumn are indeed likely to have contributed to making the military campaign more acceptable to the wider Russian public as well.

Important here is the specific political setting in Russia noted in the introduction to this book: for once the country's president and its parliament seemed to be speaking with one voice! That gave particular credibility to the security claims. The elite consensus which emerged in Russia during autumn 1999 on the severity of the Chechen threat and the necessity of a new, violent offensive against Chechnya seemed to indicate a highly surprising re-uniting among Russian politicians. Previously, the chambers of the Federal Assembly had seemed to use the formal powers they had to oppose nearly every initiative coming from the leadership.[61] The opposition had, in fact, been securitizing the Yeltsin regime itself, arguing that it was posing a threat to Russia.[62] Indeed, an impeachment process against Yeltsin had been launched by the CPRF and most political parties in the Duma in June 1999, based on the argument that Yeltsin was guilty of unleashing the First Chechen War (1994). Now the launching of a second war against Chechnya was applauded and the Federal Assembly seemed to be willing to endorse any proposals on how to counter the Chechen threat coming from the Russian leadership. When formal endorsement of emergency measures against Chechnya was needed or sought by the Russian leadership, it was duly given (see Chapter 10). This observation hints at the important question of how the power of an actor can be built through securitization processes which was raised in the theory chapter (Chapter 2) and will be addressed in more detail at the end of this book (Chapter 13).

For now, we stick to the question of how audience acceptance of the new war came about in wider circles of Russian society during autumn 1999. In the next chapter, I turn to examining the texts of 'experts' and how they interplayed with those of the Russian leadership and the political elite that autumn.

Notes

1 The Federal Assembly of Russia consists of the State Duma (the lower house) and the Federation Council (the upper house).
2 Yevgeny Primakov seems to have taken this position initially. Stating that 'we can ascertain that a sabotage terrorist war has been forced upon us', he explained the situation as a result of certain members of the security organs being connected to the criminal world ('Protiv ChP vystupayut vse', *NeGa*, 14 September 1999).
3 Lebed was quoted as saying:

As I understand it, an agreement was made with [Chechen rebel leader Shamil] Basayev, especially since he's a former KGB informant. I'm absolutely sure of this. I think Basayev and the powers that be have a pact. Their objectives coincide (…) The President and the Family have become isolated. They don't have the political power to win the elections. So, seeing the hopelessness of its situation, the Kremlin has set itself just one goal: to destabilize the situation so the elections can be called off.

When asked whether he was sure that 'the hand of power' – as he put it – was behind the [recent apartment house] bombings, Lebed replied:

I'm all but convinced of it. Any Chechen field commander set on revenge would have started blowing up generals. Or he'd have started striking Internal Affairs Ministry and Federal Security Service buildings, military stockpiles or nuclear power plants. He wouldn't have targeted ordinary, innocent people. The goal is to sow mass terror and create conditions for destabilization, so as to be able to say when the time comes, 'You shouldn't go to the polls, or you'll risk being blown up along with the ballot box.'
(Quoted in Kirill Privalov 'AND HERE'S LEBED, ON A WHITE HORSE! – Following Up on an Exclusive Interview the Krasnoyarsk Governor Gave to the Paris Newspaper Le Figaro', *Segodnya*, 30 September, 1999, p. 2)

Part of Lebed's statement was quoted in a small piece in *NeGa*, 30 September, 1999.
4 'Ruslan Khasbulatov: v Dagestane my poluchili neizbezhnoye', *NeGa*, 14 September 1999.
5 'Boris Berezovsky otvetil na obvineniya obvineniyami', *NeGa*, 17 September 1999.
6 'Protiv ChP vystupayut vse', *NeGa*, 14 September 1999.
7 'Zyuganov – za peregovory s Maskhadovym', *NeGa*, 29 October.
8 Interview with Sergey Stepashin for the programme *Geroy Dnya* ('Hero of the Day'), NTV, 5 October 1999.
9 Interview of Grigory Yavlinsky by Nikolay Svanidze for *Zerkalo* (Mirror), RTR channel, 24 October 1999.
10 Yevgeny Primakov cited in 'Ya protiv voyny v Chechne', *Trud*, 5 October 1999.
11 'Ya protiv voyny v Chechne', *Trud*, 5 October 1999.
12 'Yavlinsky ne vo vsëm soglasen s Putinym', *NeGa*, 7 October 1999.
13 'Minnats popal pod ogon' kritiki', *NeGa*, 24 September 1999.
14 Interview of Grigory Yavlinsky for the *Geroy Dnya* ('Hero of the Day') programme NTV, 28 September 1999.
15 'Declaration of Grigory Yavlinsky, head of the Yabloko faction in the State Duma or

November 9, 1999', available at www.yabloko.ru/Engl/Themes/Chechnja/yavl decl 1.html, and accessed 15 January 2016.

16 Ibid.

17 Interview of Grigory Yavlinsky by Nikolay Svanidze for *Zerkalo* ('Mirror'), RTR channel, 24 October 1999.

18 'Alexey Melnikov's responds to Anatoly Chubays with his own stringent accusations', available at www.yabloko.ru/Engl/Press/press 99nov11.html, and accessed 15 January 2016.

19 For a discussion on the absence of the Chechnya issue in the Yabloko election campaign, see H.E. Hale. 2004. Yabloko and the challenge of building a liberal party in Russia. *Europe–Asia Studies*, 56(7): 993–1020.

20 'Anatoly Kulikov schitayet chto terroristov nado bezzhalostno unichtozhat'', *NeGa*, 11 September 1999.

21 Interview with Sergey Stepashin in 'Portret bez intrigi', *Moskovskiy Komsomolets*, 18 September 1999. See also Stepashin in an interview for the programme *Geroy Dnya* ('Hero of the Day') on NTV, 5 October 1999.

22 'Terroristy proschitalis', *Vedomosti*, 23 September 1999. Aleksandr Gurov, Chair of the Committee on Security in the Duma, referred to the military operations in Chechnya as 'purging the south of Russia from international terrorist bands' ('Vzbesivshegosya zverya nado ubivat' 's, *Vek*, 12 November 1999. Translated into English by S. Mäkinen. 2008, 205, available at http://tampub.uta.fi/handle/10024/67815 and accessed 17 March 2016). 'Hotbed of armed extremism' taken from 'Minnats popal pod ogon' kritiki', *NeGa*, 23 September 1999.

23 Ryzhkov, cited in 'Vlast' i narod dolzhny ob''yedinit'sya', *Vedomosti*, 16 September 1999. Translated into English by S. Mäkinen 2008, 177.

24 For example, Chair of the Committee on Security in the Duma Viktor Ilyukhin characterized what had happened as a 'real war' and indicated there would be more terrorist acts in the future ('Protiv ChP vystupayut vse', *NeGa*, 14 September 1999).

25 For example, leader of the CPRF Gennady Zyuganov now opined that they had to draw on experience from the Second World War 'when inhabitants took turns guarding their rooftops during bombardments' (cited in 'Protiv ChP vystupayut vse', *NeGa*, 14 September 1999).

26 Zyuganov, cited in 'Dzhikhad rezhima', *Sovetskaya Rossiya*, 29 February 2000. Translated into English by S. Mäkinen 2008, 226.

27 'Iz pervykh ust. Nasha politika na Severnom Kavkaze ne mozhet byt' bol'she vyaloy', *RoGa*, 13 August 1999.

28 'Protiv ChP vystupayut vse', *NeGa*, 14 September 1999.

29 Ryzhkov, cited in 'Vlast' i narod dolzhny ob''yedinit'sya', *Vedomosti*, 16 September 1999.

30 'Protiv ChP vystupayut vse', *NeGa*, 14 September 1999.

31 This example is from Zorin's text 'My prishli v Chechnyu kak osvoboditeli', *Krasnaya Zvezda*, 20 October 1999. Yavlinsky's text from 9 November 1999 contains a similar combination of contradictory representations.

32 *RFE/RL Newsline*, 30 September 1999.

33 'Zyuganov podelil rossiyan na patriotov i predateley', *NeGa*, 1 October 1999.

34 Both quotes from 'Taynye i yavnye manëvry Moskvy', *NeGa*, 5 October 1999.

35 'Putin sozdal klub prem'yer ministrov', *NeGa*, 6 October 1999.

36 For example, Head of the Federation Council Yegor Stroyev contended: 'the Federal centre should actively engage in dialogue with all active political forces in Chechnya'. ('Rossiya ne schitayet metry', *NeGa*, 1 October 1999).

37 'Konfrontatsiya ili dialog?', *NeGa*, 28 September 1999.

38 'Skazochnik s kholodnymi glazami', *Moskovskiy Komsomolets*, 16 September 1999.

39 'Vystupleniye Putina ponravilos' senatoram', *NeGa*, 18 September 1999.

40 Cited in 'Khasavyurt byl oshibkoy?', *Trud*, 18 September 1999.

41 'My prishli v Chechnyu kak osvoboditeli', *Krasnaya Zvezda*, 20 October 1999. Translated into English by S. Mäkinen 2008.
42 *RFE/RL Newsline*, 17 September 1999.
43 *RFE/RL Newsline*, 20 September 1999.
44 'Rossiya ne schitayet metry', *NeGa*, 1 October 1999.
45 'Sovet Federatsii podderzhivayet zhëstkie mery', *Russkaya Mysl'*, 23 September 1999.
46 'Novaya Chechenskaya voyna uzhe nachalas'', *NeGa*, 21 September 1999.
47 'My razberemsya s Chechney bez pomoshchi NATO', *Argumenty i Fakty*, 8 December 1999.
48 'Kogda my nachinayem sebya uvazhat'?', *NeGa*, 14 October 1999.
49 'Protiv ChP vystupayut vse', *NeGa*, 14 September 1999.
50 'Protiv ChP vystupayut vse', *NeGa*, 14 September 1999.
51 Zyuganov, cited in 'Bespomoshchnost' praviteley kompensiruyetsya muzhestvom Russkogo soldata', *Sovetskaya Rossiya*, 1 February 2000. Translated into English by S. Mäkinen 2008, 225.
52 Zorin opined that the 'struggle against terrorism requires the forces not only of the power but the whole society: 'Iz pervykh ust. Nasha politika na Severnom Kavkaze ne mozhet byt' bol'she vyaloy', *RoGa*, 13 August 1999.
53 Ryzhkov, cited in 'Vlast' i narod dolzhny ob''yedinit'sya', *Vedomosti*, 16 September 1999.
54 'Kogda my nachinayem sebya uvazhat'?', *NeGa*, 14 October 1999.
55 Indeed, the claim that the government wanted to introduce a state of emergency in order to postpone elections was made by several people. See for example Zyuganov in 'Putin predlagayet novyy plan Chechneskogo uregulirovaniya', *NeGa*, 15 September 1999.
56 Vladimir Zorin of Nash Dom–Rossiya, in 'My prishli v Chechnyu kak osvoboditeli', *Kasnaya Zvezda*, 20 October 1999. Translated into English by S. Mäkinen 2008, 175.
57 Roman Popkovich of Nash Dom–Rossiya, in 'Terroristy proschitalis'', *Vedomosti*, 23 September 1999. Translated into English by S. Mäkinen 2008, 182.
58 Vladimir Zorin of Nash Dom–Rossiya, in 'My prishli v Chechnyu kak osvoboditeli', *Krasnaya Zvezda*, 20 October 1999.
59 Viktor Chernomyrdin of Nash Dom–Rossiya, in 'My razberemsya s Chechney bez pomoshchi NATO', *Argumenty i Fakty*, 8 December 1999. Translated into English by S. Mäkinen 2008, 180.
60 Vladimir Ryzhkov of Nash Dom–Rossiya, on Radio Svoboda, 23 November 1999, and on Radio Ekho Moskvy, 15 December 1999, quoted directly in and translated by S. Mäkinen 2008, 178.
61 As late as 13 September 1999, Communist Party leader Gennady Zyuganov stated at a press conference:

> They are preparing for emergency rule with one aim: to evade responsibility for the situation and derail elections.... There are enormous forces in the country, which are interested in fuelling the war ... the executive branch has so far commented on the events instead of taking preventive steps. The Kremlin-based party of traitors, which also exists in the Caucasus, is doing nothing to normalize the situation.
> ('Kremlin preparing for emergency rule communist leader', *Interfax*, 14 September 1999)

62 For example, in an interview with the *New York Times*, Speaker of the Federal Council Yegor Stroyev said that it would be a blessing for the country if Yeltsin left office. 'His (Yeltsin's) power does not reach further than the Kremlin walls. No one needs such a system of power. If it is preserved, we will lose Russia' ('Stroyev protiv Eltsina?', *NeGa*, 17 September 1999).

8 Expert representations of Chechnya and Russia

The point of departure of this chapter and the next is that the broader Russian public debate *could* have made a difference that autumn. Narratives that constructed the relation between Russia and Chechnya in terms different from those put forward in official and political elite language *could* have been voiced, and they *could* have spread to broader sections of the public, creating a pressure against undertaking a new war or, alternatively, halting it after some time.

If we must put a label on the Russian political system in 1999, it was still closer to a democracy than an authoritarian regime. The much-discussed installation of the Power Vertical in Russia took time, and was not in place only a few months after Putin became Prime Minister. It is reasonable to argue that the political system prevalent at that time was fairly open. And crucially, despite increasing control over media coverage of the battlefield in Dagestan, the media scene was still pluralistic during summer/autumn 1999. The fact that the three biggest TV networks (ORT, RTR and NTV) invited politicians as different as Prime Minister Vladimir Putin, Governor Aleksandr Lebed and the liberal opposition politician Grigory Yavlinsky to comment on the government's handling of the situation in Dagestan and the bombings in Russian cities testifies to this.[1] The imposition of a media blockade on Chechnya was a gradual process.[2]

Counter-securitizing or de-securitizing attempts *could* have been launched at this stage. If such alternative discourses had found strong resonance with the Russian public, it would have made it difficult to undertake a new war against Chechnya legitimately. Again, developments during the first war in Chechnya (1994–1996) are instructive. The securitizing narrative offered by the Russian government at the time, which represented the Chechen regime as 'criminal' and focused on the need to protect Russian territorial integrity, was not well argued at the outset. Moreover, Mickiewicz (1997) contends that the Yeltsin regime proved unable to 'manage' the war as a daily discursive event for which it had to compete with other sources of information.[3] The sharp discursive struggle which emerged as the First Chechen War unfolded between the official discourse and an alternative discourse which served to de-securitize Chechnya/Chechens was commented on in Chapter 6. This discursive struggle finally resulted in the Russian public rejecting the official securitizing narrative.[4] Such a weak foundation for the policy of war among the Russian public during the First Chechen War was undoubtedly one factor that pushed Russian authorities toward ending hostilities and deciding to sit down at the negotiating table in 1996.[5] This time the situation was different.

In what follows below and in the next chapter, expert opinion pieces and journalistic accounts on Russia and Chechnya from autumn 1999 are analysed. It might be objected that the selection of these groups as representing the Russian public is not satisfactory – numerically, experts and journalists make up a very small part of the Russian public. However, they are quite influential compared to their size, in terms of mediating and weighing in on discursive struggles and shaping the public debate on key issues – particularly, it seems likely, on an issue such as a counterterrorist operation, so physically and mentally distant from the daily life of the 'man in the street'. Moreover, the choice of expert and journalistic texts as representatives of the 'Russian public' is highly satisfactory in terms of methodology. Such texts give direct access to linguistic representations, essential to discourse analysis. By contrast, public opinion polls or interviews at a later point in time are less reliable sources, as they give only indirect access to representations and are often mediated through questions that necessarily involve some kind of bias.

Let us first see what expert representations of Chechnya and Russia in Russian newspapers looked like during autumn 1999. I identify three positions here: one stronger, in terms of the number of opinion pieces that can roughly be categorized within this position, which resonates strongly with official representations of Chechnya and Russia that autumn; another, much weaker, which can be identified as the remnants of the interwar 'discourse of reconciliation'. This second position allows for more nuanced and less radical representations of Chechnya and Russia and also suggests less radical emergency measures than the dominant position does. Finally, I find a middle position: it seems to originate in the 'discourse of reconciliation', but goes a long way towards accommodating the new official representations of Chechnya.

While exploring the content of these expert texts is an aim itself, a core exercise again involves comparing them against the 1999 official securitizing narrative presented in Chapter 5. Given the conceptualization of securitization as an intersubjective endeavour, it is relevant to see *how* expert accounts feed into and contribute to the emerging discursive consensus on Chechnya as an existential terrorist threat. While the texts of the political elite in the Federal Assembly contribute to this construction in a similar fashion as official texts, expert texts exhibit a different style and invoke different references in seeking to build authority around their arguments.

Chechnya as a lawless and violent space

Expert representations of Chechnya and Russia that autumn were more varied than representations among the political elite. That said, representations in most expert texts add up to a narrative that echoes the official securitizing narrative.

As we will see below, the sum of representations emerging from various newspaper opinion pieces and editorials construct 'Chechnya' as an entirely lawless, violent space.[6] It is represented as a place where 'guns are the main labour units'[7] as 'the one and only terrorist state in the world – a hotbed of instability for the entire Caucasus'[8] – or as 'an abyss of anarchy and immorality ... everywhere there is injustice, lawlessness and chaos'.[9] Even if this discourse on Chechnya as a lawless and violent place does not always add up to an explicit argument for war such as that found in official language during autumn 1999, it serves to underscore the official securitizing moves. Expanding on and elaborating the image of Chechnya as lawless and violent and endowing this representation with expert authority adds to the distance already constructed between 'Chechnya' and 'Russia' in official language.

Moreover, a set of expert opinion pieces and editorials also include language that constructs Chechnya as a direct *threat* to Russia. Here the Chechen threat is characterized with descriptors such as 'killers and terrorists',[10] 'throat-cutters',[11] 'rude, bearded, ruthless bandits',[12] 'criminals' and 'wolves',[13] 'pathological murderers ... having carried out ethnic cleansing ... committed massive crimes',[14] 'criminals, drug addicts and bandits ... thoughtless killers, heartless and capable of murdering their brother, sisters, mothers and fathers'.[15] Also recurrent are references to the 'Chechen Wahhabis' as dangerous and motivated by money, not by the Muslim faith.[16] Returning to the idea that threat representations can be placed on a scale, even this rough enumeration shows that representations in many Russian expert texts during autumn 1999 can be placed at the top end of the scale, as *existential threats*.

On this account, they are not unlike the representations found in official texts and in those of most of the political elite. Quite instructive is an opinion piece by political scientist (*Politolog*) Viktor Gushchin titled 'Terrorism is a psychological war', printed shortly after the apartment bombings and shortly before the launching of the Second Chechen War. The opinion piece illustrates how similar 'expert' language was to 'political' language. While stressing that the resolution of any crisis has to start with acknowledgement of 'facts' and thus elevating the text to an authoritative level of expert objectivity and truth, Gushchin's account of the situation of Russia is as emotional, stark and terrifying as any of the political texts referred to in the previous chapter. Second, it carries within it several characterizations of the threat that can be found in other expert texts that autumn, and is therefore fairly representative of expert language on Chechnya. Gushchin presents these 'facts':

For terrorists nothing is forbidden, impossible or inadmissible. Those who are faced with a terrorist war need to understand that it does not lead to life, but to death. In the eyes of the terrorist, fear before death should be total. No one should be excluded, not old people, or children, or women. Terrorist war is a war of destruction, but of a special kind. Terror is foremost a war to destroy human dignity. It is destruction by fear ... the terrorists intend to go to the bitter end and the explosions will continue as inevitably as the sun rises and sets.... The initiators of the terrorist war will never and for nothing decline from their goals and intentions.... They will either succeed in overthrowing Russian power by the means of our hands or everyone down to the last man will die. This is the objective logic of the psychological terrorist war. It is wrong to suggest that the reason behind the war is retaliation for some local defeat ... the core of this psychological terrorist war lies way back in time. We have long since forgotten to pay attention to the fact that the organizers of this terrorist war against Russia have continued the war for hundreds of years. The victory over Russia, they claim, is not merely a historical duty, but a genetic duty, like David's victory over Goliath. We have only one choice: either to meet this challenge or to die.[17]

Here Chechnya is not specifically mentioned, but the threat is explicitly linked to Russian history and the asymmetric and violent relation between 'brothers', which alludes to Russia's encounter with the Caucasus. The threat facing Russia is presented as continuous and unchangeable (even genetic) – an idea repeated in several other expert accounts as well.[18] Moreover, the threat is presented as lethal, inhumane and overwhelming; and the gravity of the situation is underscored by repeated references to the situation as 'war'. The idea that a 'war' has been launched against Russia re-appears in many expert texts during autumn 1999 in expressions such as 'terrorist war against peaceful population'[19] or 'massive terrorist war has been declared on Russia'.[20] As in the official narrative and political elite representations, parallels are drawn to the Second World War. Some expert texts go even further and indicate a direct equation to Nazism, as in the opinion piece by Sergey Roy, Editor-in-Chief of *Moscow News*, on the pages of *Nezavisimaya Gazeta*, where he writes that 'Chechen "Nazism" needs to be repressed, Nazism is a threat anywhere in the world'.[21]

As shown by the latter quote, the tendency identified in political elite discourse of conflating not only 'Chechnya' but also 'Chechen' or 'North Caucasian' with 'terrorism' or other concepts of threat is present in expert language as well.[22] Although not widespread, the related idea that the Chechens are collectively guilty and therefore deserve punishment is at times quite explicit. For instance, the philosopher Zemlyanoy writes:

every nation deserves its leaders.... When Maskhadov, instead of arresting the field commanders and fighters responsible for terror, theft and kidnapping, appoints them as commanders in the 'holy war' against Russia, the Chechen nation, or the part of it which has not fled, support him. The Chechen leader applies to his countrymen not only collective responsibility, but also collective guilt.[23]

Other expert commentaries make this equation between leaders and nation even more explicit. Sergey Roy, for example: 'The Chechens as a nation can all be considered guilty since they have accepted Basayev, a pathological murder, as their hero and therefore deserve a "massive punishment" equivalent to the "massive crimes" they have committed.'[24]

We see that Russian expert representations sometimes go further than official representations in terms of portraying 'Chechens' as guilty, thus contributing to legitimizing violent retribution not only against the Chechen regime or the Chechen militants, but also against the Chechens as an ethnic group.[25] Official language on Chechnya seldom moved from securitizing Chechnya as a terrorist threat to representing the Chechen ethnic group as dangerous or guilty.

Official and expert representations are more congruent along other dimensions. Viewing representations of 'Maskhadov' as an event within the event, we find that he is not mentioned very frequently in most expert texts. However, when he is mentioned, he is stripped of his legitimacy as the president of Chechnya, through expressions such as 'the so-called President Maskhadov'[26] or claims that the elections that made him president in 1997 were 'un-constitutional'.[27] Maskhadov is only once represented as a 'killer and terrorist',[28] but he is frequently identified as consenting, as part of the dangerous Other.[29] The idea of Maskhadov as 'illegitimate' in legal terms and that of him as 'an accomplice' both parallel those found in official language during autumn 1999 (Chapter 5).

Moreover, as in the official narrative the level of danger implied in expert representations of Chechnya is enhanced by recurrent references to 'a distant enemy'. These constructions of related but distant forces which stand behind and nourish the Chechen threat make the local conflict look like one of proxy, while simultaneously increasing the magnitude of the frontier.

Two such forces can be identified in expert texts. First, as in official representations, Middle Eastern countries or Islamist movements or organizations are pointed out. For example, the opinion piece written by doctor of law Ramzan Dzhabarov entitled 'Extremists against traditionalists, the Islamic Factor in Chechnya and its foreign sponsors' constructs the Wahhabi strand of Islam emanating from Saudi Arabia as the Mastermind behind the Chechen threat and the true 'distant enemy' harbouring the ultimate aim of subverting Russia. In this account, both the Chechen warlords and the Chechen regime have been co-opted by the 'dangerous Wahhabis'.[30] Other expert accounts identify a possible connection between Bin Laden and Chechnya, linking together the Afghan and Chechen 'terrorists'. Authority is given to this claim by referring to the early recognition of these links by U.S. and Israeli special forces and the presentation of documents by Russian authorities proving this fact.[31]

Second, several expert texts point to the U.S./NATO as a force behind the threat facing Russia in the North Caucasus. For example, Anatoly Kucherena writes:

> it is easy to imagine that huge financial flows will be directed toward the region, also military and all kinds of specialists will enter the region from the entire world. After that the USA with full right could claim the North Caucasus to be a zone of American national interest. What other than a chance to strengthen NATO southern flank would it be?[32]

In constructing the Chechen threat as simply an offspring of the eternal U.S. threat, these texts often suggest that any criticism of Russian policies toward Chechnya is a tool for harming Russia. This representation, then, is quite similar to that identified in several political elite texts, but which was non-existent in official language at the time.

A curious twist, and actually quite a widespread one, is the merging of these two 'distant enemies' into one. We find a clear example of this in Leonid Ivashov's opinion piece, where the threat facing Russia in the Northern Caucasus is equated with that plaguing Tajikistan and Kyrgyzstan. The common source behind these security threats is initially identified as Osama bin Laden, but NATO and the USA emerge as the greater threat standing behind him.[33]

Taken together, the level of threat implied in these expert representations of Chechnya amount to it being 'existential'. This is underscored by references to the situation as a 'war' fostered by diverse distant enemies. As a next logical element in the narrative, most expert texts communicate a sense that Russia now finds itself at the *point of no return*. The future that the country would face if 'Russia a second time round shamefully stands aside in the face of rude, bearded, ruthless bandits' is indicated as 'the collapse of the unity of the Russian state'[34] or as Russia 'sinking to the bottom'[35] as 'totally discrediting the Russian state power's ability to be the master in its own house'.[36] Sometimes the situation is given in terms of an ultimatum: 'We have only one choice: either to meet this challenge or to die'.[37]

The way out

Given this sense of urgency and danger, it is not surprising that the emergency measures proposed in these expert texts as the possible *way out* are often as radical as those in the official narrative.[38] According to Sergey Roy, for example, the situation calls for 'massive punishment', including a 'total blockade' around Chechnya. Not a 'caricature of a blockade', but the establishment of :

> a proper military front line all around Chechnya. From this line, using regular forces, one Chechen village after another should be suppressed, by a total filtration of the population and by subjecting every fighter, however many they might be, to a military field court.

Roy advises carrying out special operations to take out the 'fifth column' in the regions surrounding Chechnya, as well as in the heart of Russia and particularly in Moscow; stopping all pro-Chechen propaganda in Russian media outlets; creating a cordon sanitaire which would destroy the system of channels financing and arming the 'Chechen terrorists'; and that all other types of political, diplomatic, economic support from foreign governments should also be stopped.[39] In short, this particular opinion piece sanctions the official suggestions of emergency measures towards Chechnya in full. It even functions as a kind of foreboding that the 'cleansing operations' that became so widespread in Chechnya during the Second Chechen War would be necessary and just.

Other expert texts also indicate the need for radical emergency measures, even representing them as 'humane' in the given situation:

> The war against terrorism is the fight of evil against evil, in its absolute expression. Justice and humanity is not to reject the radical measures of the war against terrorism, but to use them.... The only way to prevail over terrorism is by employing the methods of terrorism, not only in the territorial field, but also in the psychological field ... the authorities are obliged to demonstrate confidence in their own strength and really resolve to fight terrorism by any methods and means, including retaliation.[40]

The flipside of representing this harsh, violent retribution against Chechnya as necessary and just is the rejection of measures that encompass contact, compromise and negotiation with Chechnya. For example, with reference to 'the conduct of the Chechen power', Professor Vadim Pechenev notes 'it would be fatal to "reconcile" or "sweeten" Grozny'.[41] The logic in this and similar constructions is based on the congruence between, on the one hand, the level of danger implied in the threat representations and on the other, the measures suggested. Sergey Zemlyanoy's reasoning demonstrates this correspondence nicely: 'Because Chechnya is in the hands of the "field commanders", or rather incorrigible throat-cutters and qualified terrorists, there is no real partner for the federal power to negotiate the political solution to the Chechen problem with.'[42]

In the dominant variant of the political elite discourse on Chechnya, voices advocating policies of reconciliation were often represented as naïve and dangerous. This is an argument we find in expert texts as well. Viktor Gushchin characterized as 'stupid' the argument that:

> Russia must never degrade itself to the level of the terrorists by adopting their ways and never act according to the principle of 'an ear for an ear and a tooth for a tooth'.... In relation to the war on terror opinions such as these cannot be seen as anything other than capitulation encouraging continued terror against defenceless people.[43]

In other expert accounts, Russian liberals who urge negotiations with Maskhadov are characterized as 'infantile humanists'.[44] Again this is a representation not found in official discourse during autumn 1999. Describing critics of the war on terror in Chechnya as 'dangerous' appeared first in expert texts and in the texts of the political elite in the Federation Assembly. Only later did it become a central theme in official language.[45]

Expert texts chimed in with official language during autumn 1999 on the more general call for 'unity' as a means of withstanding the terrorist threat, however. First, many warn that lack of unity or even any opposition within the Russian entity could be harmful in the situation. These texts urge a stop to 'intrigues and competition',[46] 'fifth columnists' and 'enemies within'.[47] Not doing this would 'be dangerous',[48] 'play into the hands of the terrorists'[49] or bring Russia to 'chaos'.[50] Second, in several expert texts the crisis facing Russia is presented as an opportunity for moving out of the divisive and chaotic 1990s and into creating a new unity and a stronger Russian state.[51] In the words of Aleksey Podberezkin, the crisis has 'awakened people's feeling of pride and self-esteem.... We have to use this chance to unite'.[52] Third, unity in a strong state is given as Russia's natural and right state of being. Doctor of Philosophy Sergey Zemlyanoy notes that historically a 'moral catharsis had been brought to Russia with terrible sorrow and big distress'. In this situation 'Basayev, Khattab and their throat-cutters' could serve to unite all branches of Russian power and finally direct efforts toward securing the common interests of the Russian state. According to Zemlyanoy, the 'state sense' had been awakened with Putin, who was working against the anti-state policies of the Yeltsin regime to make Russia strong.[53]

Sergey Kazennov and Vladimir Kumachev, who belong to influential institutions in Russian academia, employ a different language, but in essence their representation of 'unity' and 'strong state' as the right organizing principles for Russia is the same. With reference to the nineteenth-century philosopher Berdyayev, who wrote that 'for Russia there is no such thing as individual salvation, we can only be saved together', the authors lament the fact that there are no common values in Russia. The North Caucasus is represented as merely an extreme and brutal expression of this general problem. Their suggestion for a Russian common platform includes statist and social elements, as well as elements of law and order and an emphasis on morals, ethics and conservative values.[54]

To sum up, in what I have identified here as the dominant position within expert discourse, the suggested 'way out' fits the official narrative and policy proscriptions fairly well. Experts even argue that this violent and uncompromising approach is 'humane', and that proponents of non-violent measures such as communication and negotiation are 'stupid' and 'infantile'. The call for *unity* as a means of withstanding the terrorist threat also parallels that found in the official narrative (Chapter 5). Moreover, *unity* is given credibility as Russia's natural and correct state of being by expert references to Russian history and philosophy. And with this, we have moved into the question of how *Russian identity* is re-articulated in the process of defining the Chechen threat.

Russia as united, strong and orderly

As the above quotes on 'unity' show, 'strong' is also dominant in expert representations of what Russia is, or should be, in encountering the Chechen threat. Several expert opinion pieces point to Russia's weakness as the main reason for the security problems facing the country,[55] indicating 'the Russian state power's ability to be the master in its own house' or similar ideas of strength as the solution.[56] Simultaneously, the official argument of Chechnya as the 'offender' and Russia the 'defender' is echoed in the dominant expert position. Also the related notion of Russia as 'innocent' can be found in expert representations in expressions such as 'they plant explosives in our homes and we turn the other cheek'.[57] Indeed, Putin's phrase 'we need to throw off all our complexes, also our complex of guilt', finds a direct parallel in Roy's 'we have to stop being ashamed of what Russia is: a pseudo-democratic empire. A minority should not be able to threaten a majority'.[58]

The most widespread juxtaposition in expert texts, however, and one that attaches a moral superiority to this Russian majority, is that representing Chechnya as 'chaos' and 'lawlessness' and Russia as 'order' and 'civilization'. This more traditional classification of the Caucasus as the opposite of Russian 'civilization' is at times very explicit, as in Sergey Zemlyanoy's statement:

> The mountains give birth to more people than they can feed; civilized modern laws don't work in the mountains. The only way to withstand these ancient criminal ways (*stikhiya*) is to introduce colonial administration, supported by armed force.... There are no proper authorities in Chechnya, just armed formations based on clans, the right of the strongest and sometimes structured by *Sharia* and *tarikat* ... a society that has regressed to its eternal pattern.[59]

Other accounts include the historical Caucasian opponent on the side of 'order' with Russia, something which constructs the Otherness of present Chechnya as even more radical. Vladimir Degoyev, for example, argues that there are few lessons to learn from studying the Caucasian wars for advice on how to deal with present-day Chechnya. While Imam Shamil is represented as a respectable and honest person and his staunch opposition to Russian colonization in the nineteenth century as an honourable undertaking, the current Chechen leadership is given the opposite identity. According to Degoyev, Dzhokhar Dudayev decided to 'try out Chechen national sovereignty with a dictatorial technocratic leaning'. While 'Shamil turned "paternalistic chaos" into "Islamic order", the present Chechen reformers have turned "Soviet order" into "Islamic chaos"'.[60]

Malik Saydullayev's appeal to the Chechen nation, posted on the front page of *RoGa* on 14 October 1999, draws a similarly sharp line between today's Chechnya and the 'true' Chechnya:

> during the past seven years, and at the will of enemies, the nation had been drawn away from normal human life and thrown into the abyss of anarchy and immorality.... In these past years Chechnya changed from being the most developed republic of the Northern Caucasus into the most backward and poor region ... everywhere there is injustice, lawlessness and chaos.... They tell you that you are living in a free country, but the only thing you are free from is laws, both human and those of God...) [Chechens must stand together to] punish the bandits, chastise the torturers and destroy the killers [and become] good neighbours.[61]

We see, then, that in expert discourse also the identification of the Chechen threat serves the function of re-articulating Russian identity, as in official language. Characteristics such as 'unity', 'strength', 'innocence', 'order', 'laws' and 'civilization' are presented as key features of 'Russia' juxtaposed to 'Chechnya'. This re-articulation of Russian identity is strikingly similar to that found in official language during autumn 1999 (Chapter 5). That in turn indicates that this position on what Russia is, or ought to be, is intersubjectively constructed – not something proposed from the top political level and then adopted into expert language.

We have seen that this is the case with representations of 'Chechnya' as an existential threat in most expert texts as well. Although expert texts may add broader dimensions to the construction of 'Chechnya' with more elaborate accounts of chaos and lawlessness and heavier historical and legal references, they also offer representations of 'Chechnya' as an existential threat that are similar to, but yet independent of, official representations. We can conclude that the position which combines the discourse on Chechnya as an existential terrorist threat with that of a strong, orderly and innocent Russia was alive in Russian expert language and emerged onto the pages of newspaper opinion pieces during autumn 1999. From there it met and merged with official language and the language of the political elite to harden and then overwhelm alternative positions on Chechnya and Russia.

Alternative expert positions

This is not to say that the position that dominated the official interwar discourse, here termed 'the discourse of reconciliation', was totally absent from expert language. It could be found in smaller publications with lower circulation and was voiced by liberal thinkers such as Anatoly Pristavkin and Yury Afanasyev. In their narrative, the Chechens were victims, just as Russian soldiers were; the Russian regime was guilty of the Chechen tragedy, its policies comparable to those of the Hitler regime – and the way out was to avoid new bloodshed, simply because it would lead to even more bloodshed.[62] This way of representing the Russo-Chechen relation also continued to dominate the language of 'practising expert groups' such as the Committee of Soldiers' Mothers. At their press conference in September 1999, they blamed Russia's politicians for the problems in Chechnya, declaring: 'Yet again the politicians attempt to solve their problems with the blood and lives of our children'.[63] Unlike the case in the First Chechen War, however, their talk never made it into the opinion piece pages.

Here we can note one exception among the *NeGa* and *RoGa* expert opinion pieces: that by political scientist Vadim Belotserkovsky titled 'It is hard not to believe Basayev'.[64] This text turns the identification of Chechnya and the Russian regime on its head by constructing the latter as an existential threat to the 'smaller nation'. By depicting Russia as no less criminal than Chechnya, the Russian regime as manipulative and propagandistic, and concluding that 'Russian medieval morals demonstrate the absence of humanism', this opinion piece in effect places Russia on the side of 'barbarians' and not Chechnya.

A few other expert opinion pieces can be placed in a middle position in terms of the degree of threat and Otherness they attach to Chechnya. In these texts, 'Chechnya' is usually linked to the 'Arab East', 'terrorism' or 'Islamic movements': on the other hand, several of them also note the diversity in Islamist organizations, attaching very different degrees of danger to them.[65] There are also explicit warnings against equating Caucasians with terrorists, against making the Caucasians the culprit of everything and in the final event splitting Russia along ethnic lines.[66] The text, written by the well-known ethno-anthropologist Valery Tishkov, draws attention to the grievances of the Chechen civilian population and highlights the relation of interdependence between Russians and Chechens as fellow citizens.[67] In this group of texts, 'Chechnya' is constructed in a less monolithic way, and with a lower degree of danger, than in 1999 official texts. However, they do not overlap with representations of Russia in the 'discourse of reconciliation', as they stop short of blaming and shaming Russia.

What then of the 'way out' indicated in these texts? Here we find that it fits logically with the less radical construction of Chechnya. Historians Konstantin Polyakov and Akhmat Khasyanov, for instance, normalize new Russian practices undertaken as a response to 'terrorism' by showing the parallel to practices undertaken in the Middle East such as the 'establishment of cordons sanitaires' or 'harsh military response' – but they also warn explicitly against relying solely on the use of force to solve problems of terrorism, and indicate the need to address the root causes in the social, economic and political fields.[68] Similarly, political analyst Aleksandr Sabov links 'Chechnya' to terrorism, Islamism and violent conflict in Algeria, but also emphasizes the many different strands of Islamism, and argues that negotiation with the Islamists is the only way out of conflict.[69] The text by Tishkov, which underlines the affinity between the Russian and the Chechen civilian populations, explicitly states that the key 'technology to solve the Chechen tragedy is teamwork and contact between the conflicting parties'.[70]

As well as nuancing and at times contradicting the image of 'Chechnya' in official representations, this middle position ensures that policies other than the use of violence are kept alive in the public mind as a possible way of dealing with Chechnya. At the same time this position seems to accommodate official representations, in that it accepts locating Chechnya within the orbit of 'Islamism' and 'terrorism', and seems to promote non-violent measures as a supplement, not an alternative, to the forceful measures indicated in official discourse. This type of accommodation of official language in expert texts parallels that found in the political elite texts (Chapter 7).

Conclusion

Looking at these expert texts as a whole, we find that a middle position is more strongly articulated than the 'discourse of reconciliation', which seems to have retreated to smaller and more marginalized outlets. But even this middle position accommodates the 1999 official narrative to such an extent that it can hardly be considered a competing discourse. Moreover, it is much less prominently mobilized than the position that I have termed the 'discourse of war'. The position that one-sidedly represents 'Chechnya' as an existential threat, and demands violent retribution, drowns other positions on the opinion piece pages of *NeGa* and *RoGa* in autumn 1999.

The striking degree of overlap between representations of threat and referent object found in expert texts and those in official texts indicates *acceptance* in this part of the Russian audience for launching a second war against Chechnya. Even more important to this study is how most expert representations interacted with and enhanced official representations to produce a dominant version of the kind of challenge constituted by Chechnya. This process was intersubjective in nature, as the audience – the Russian expert community – contributed to the 'outcome'. They were in no way silent recipients of a pro-war narrative constructed solely from the top of the political system. While several political elite texts had merged the Chechens/Caucasians under the 'terrorist' label, certain expert texts spoke of the 'collective guilt' of the Chechen people in a way that official representations never did. This served to merge the Chechens as people into the terrorist threat and eventually to legitimize the use of violent measures against them as a group.

The investigation of expert texts also showed that most of them gave the use of tough, violent measures against Chechnya as the only possible *way out*. Some even seemed to indicate that more radical measures than those suggested in the official narrative were necessary. A notable input from political elite and expert text alike was the idea that negotiation or contact with the Chechen enemy was dangerous, and even that those who advised such policies in Russia were dangerous. These ideas were to be voiced by the Russian leadership later on, but not during autumn 1999.

The notions of *unity* and *strong state* have a special status in most expert texts, as in official texts. These are given as preconditions for withstanding the Chechen threat, while disagreement and division of power is given as dangerous. Moreover, in many expert texts unity is represented as the primordial and 'true' state of being for Russia, with the new war as an opportunity to break with the chaos of the 1990s and return to this natural state of being. The Russian leadership really struck a chord in the expert community in elevating unity and strong state as the core elements of Russian identity.

On the other hand, it is also noteworthy how the *style/genre* employed in the expert texts helped authorize and expand the official securitizing narrative. Many expert accounts included longer historical perspectives and philosophical explications that served to underscore official representations on Chechnya or to supplement them, as with the reasoning on the collective guilt of the Chechens or Russian unity and strength.

Finally, this review has shown that the language used by experts, leaders and the political elite in Russia were at times very similar. Several expert texts employ language that is heavily emotional, strong and one-sided. This might say something about the politicized nature of the Russian expert community, but the key point here is the way in which some expert texts function as a direct echo of official language. They give credibility to the claim that Chechnya is an existential terrorist threat, through a simple logic of repetition. Repeated iterations make the claim more believable, particularly when the writer is an expert.

Notes

1 'Telebitva za golosa izbirateley nachalas', *NeGa*, 21 September 1999.
2 J. Wilhelmsen 2003. *Norms: The forgotten factor in Russian–Western rapprochement: A case study of freedom of the press under Putin.* FFI Report 00457. Kjeller: FFI.
3 E. Mickiewicz. 1997. *Changing Channels: Television and the struggle for power in Russia.* New York: Oxford University Press.
4 Wagnsson (2000, 179) refers to polls indicating that 'the alternative of letting Chechnya leave the Russian Federation was rather acceptable to the public mind in 1996, and became increasingly accepted with the passing of time' and notes that 'only a tiny minority, only 5 per cent, believed that the territorial integrity was important enough to justify armed actions in Chechnya'. See C. Wagnsson. 2000. *Russian Political Language and Public Opinion on the West, NATO and Chechnya.* Stockholm: Akademitryck AB Edsbruk.
5 See, for example, M. Evangelista. 2002. *The Chechen Wars. Will Russia go the way of the Soviet Union?* Washington, DC: Brookings Institution Press, 42.
6 See, for example, Professor Vadim Pechenev (PhD in Philosophy) 'Kamo gryadeshi?', *NeGa*, 13 October 1999.
7 Col. Aleksandr Veklich 'Dvoynaya moral' Aslana Maskhadova', *RoGa*, 13 October 1999.
8 Anatoly Kucherena (lawyer) 'Vooruzhennyy myatezh kak istochnik prava?', *NeGa*, 29 September 1999.
9 Malik Saydullayev (Chairman of Humanitarian Help to Chechnya), 'Maskhadov stal poslushnoy igrushkoy v rukakh otpetykh banditov i inostrannykh khozyayev', front page of *RoGa*, 14 October 1999.
10 Sergey Roy (Editor-in-Chief, *Moscow News*) 'Smertel'naya ugroza', *NeGa*, 17 September 1999.
11 Sergey Zemlyanoy (PhD in Philosophy (*Kandidat*) 'Politicheskaya situatsiya v Rossii', *NeGa Stsenarii*, 13 October 1999.
12 Ibid.
13 Sergey Artemov (Lieutenant Colonel) 'Vozhaki volch'ih stay', *RoGa*, 22 October 1999.
14 Roy 1999.
15 Saydullayev 1999.
16 Ramzan Dzhabarov (doctor of law (*Kandidat*)) for example characterizes them as 'normal terrorists, for whom the Islamic doctrines are not important, only how they pay off'. ('Ekstremisty protiv traditsionalistov', *NeGa*, 20 October 1999). Likewise, Malik Saydullayev constructs the Wahhabis as godless, warning his fellow Chechens: 'They tell you that you are living in a free country, but the only thing you are free from is laws, both human and those of God' (Saydullayev 1999).
17 Viktor Gushchin (Political Scientist (*Politolog*)) 'Terrorizm – eto voyna psikhologicheskaya', *NeGa*, 21 September 1999.
18 'Chechnya cannot part with Russia but it will always remain a threat to the security and integrity of Russia', Roy 1999.
19 Aleksey Podberezkin (leader of VOPD 'Dukhovnoye Naslediye') 'Rossiyskiy krizis i krizis oppozitsii', *NeGa*, 22 September 1999.

20 Gushchin 1999.
21 Roy 1999.
22 Headlines such as 'Caucasian explosions' for example serve to attach danger to Caucasians (Pechenev 1999).
23 Zemlyanoy 1999.
24 Roy 1999.
25 Accounts of the 'good Chechens' within this position are fairly slim, but appear in the juxtaposition of 'violent Wahhabis' with 'peaceful Sufis' in Chechnya. While the Wahhabis are presented as creating chaos and being a threat to the state, the official clergy are represented as creating 'order', they are said to 'take an active part in re-establishing the Chechen state and leadership of the country, creating an atmosphere of stability in society' ('Ekstremisty protiv traditsionalistov', *NeGa*, 20 October 1999). This construction has become quite widespread in Russian discourse in later years and is a core juxtaposition in the official discourse of the Ramzan Kadyrov regime.
26 Saydullayev 1999.
27 Roy 1999.
28 Ibid.
29 'Maskhadov is not capable of seeing that he is a marionette for Basayev, Khattab, Raduyev and other incorrigibles, who use him as a "roof"' (Zemlyanoy 1999). 'He became the loyal toy in the hands of irreparable bandits and foreign masters' (Saydullayev 1999).
30 Dzhabarov 1999.
31 Konstantin Truevtsev (Associate Dean of Applied Politics Department at the Higher School of Economics (Vysshaya Shkola Ekonomiki)) 'Ben Laden v kontekste Chechni', *NeGa*, 30 November 1999.
32 Kucherena 1999. Sergey Zemlyanoy argues in the same vein:

> Chechnya and Northern Caucasus on the whole have become a battlefield for geopolitical forces aiming to pull Russia apart.... Under the dictate of the USA and NATO, Russia is increasingly becoming an object instead of a sovereign subject in international politics, and if this encroachment on Russia's sovereignty doesn't stop, the face of Russia will disappear.
>
> (Zemlyanoy 1999)

33 Leonid Ivashov (Head of Department for international cooperation, Russian Ministry of Defence) 'Rol' Rossii v uregulirovanii konfliktov usilivayetsya', *NeGa*, 18 September 1999. See also Truevtsev 1999.
34 Zemlyanoy 1999.
35 Roy 1999.
36 Pechenev 1999.
37 Gushchin 1999.
38 Language parallel to that found in official statements is at times quite striking in expressions such as 'terrorists need to be exterminated. There is no place for them on earth' (Veklich, 1999). Or Saydullayev's advice to 'punish the bandits, chastise the torturers and destroy the killers' (Saydullayev 1999).
39 Roy 1999. Pechenev also argues for the 'use of different defensive forceful measures, even preventive large-scale operations in all regions where there is active criminal activity in Northern Caucasus' (Pechenev 1999).
40 Gushchin 1999.
41 Pechenev 1999. A similar message is conveyed by 'No one and nothing can appease terrorists' (Gushchin 1999) or 'The danger lies not in conducting a full-scale ground offensive, but in not pursuing it to the end' (Zemlyanoy 1999).
42 Zemlyanoy 1999.
43 Gushchin 1999.

44 Zemlyanoy 1999.
45 For example, Vladimir Putin in his address on 21 November 2007 referred to what he termed 'jackals' in Russia:

> those who, in the most difficult moment, during the terrorist intervention into Russia [from Chechnya], treacherously called for negotiations, in fact for collusion with terrorists, with those who killed our children and women, speculating in the most unscrupulous and cynical way on the victims. In short, these are all those who, towards the end of the past century, led Russia to mass poverty, [and] ubiquitous bribe taking.
> (Cited in 'Putin's "Jackals"', *Wall Street Journal*, 30 November 2007)

46 1999.
47 Gushchin 1999.
48 Pechenev 1999.
49 Gushchin 1999.
50 Podberezkin, 1999.
51 Sergey Kazennov and Vladimir Kumachev describe the lack of a common ideology to guide Russia in the 1990s, indicating the lack of such an ideology has made Russia 'weak' and therefore 'an object of expansion from the outside and from the inside' (Sergey Kazennov (Head of Department geostrategic studies IMEMO) and Vladimir Kumachev (Vice president, Institute of National Security and Strategic Studies) 'Umirit', a ne usmirit', *NeGa*, 22 September 1999). Sergey Zemlyanoy also frames Russia's situation in autumn 1999 as one of total moral decay (Zemlyanoy 1999). Pechenev (1999) points to the need for all-Russian unity on how to solve the North Caucasian problem and to strengthen the Russian state.
52 Podberezkin 1999.
53 Zemlyanoy 1999.
54 Kazennov and Kumachev 1999.
55 See for example Zemlyanoy's argument that 'the weak are beaten, the weak are not reckoned with, the weak are looked down upon' (Zemlyanoy 1999).
56 Pechenev 1999. See also Ivashov 1999.
57 Zemlyanoy 1999.
58 Roy 1999.
59 Zemlyanoy 1999.
60 Vladimir Degoyev (D.Sc. (History) Professor, MGIMO) 'Chechenskaya voyna: Retsidiv ili fenomen?', *NeGa*, 17 September 1999. Degoyev's representation of Dudayev's government is similar to that of Anatoly Kucherena, who described it as 'a period of total violence, impunity and lawlessness' (Kucherena 1999).
61 Saydullayev 1999.
62 Interviewed in *Obshchaya Gazeta*, 30 September 1999.
63 Small note quoting the press conference 'Opredelën poryadok voyennoy sluzhby', *NeGa*, 21 September 1999.
64 'Mne trudno ne verit' Basayevu', *NeGa*, 17 November 1999.
65 Konstantin Polyakov and Akhmat Khasyanov (*NeGa*, 7 October 1999), Aleksandr Sabov ('Generaly s imamami ishchut obshchiy yazyk', *RoGa*, 12 October 1999).
66 Dzhafar Sadyg (journalist), 'Nuzhen poryadok, a ne proizvol', *NeGa*, 25 September 1999. Pechenev 1999.
67 Valery Tishkov (Director of the Institute of Ethnology and Anthropology, RAN) 'Kak likvidirovat' katastrofu', *NeGa*, 12 August 1999.
68 Polyakov and Khasyanov 1999.
69 'Generaly s imamami ishchut obshchiy yazyk', *RoGa*, 13 October 1999.
70 Tishkov 1999.

9 Journalistic representations of Chechnya and Russia

In securitization for war, journalistic texts are decisive. If representations in journalistic accounts differ sharply from those of the official narrative, that can be taken as a sign of non-acceptance by this group. Whether journalists deem the new war a legitimate undertaking or not can be read from their accounts. But these accounts also have a wider and more important function: creating a link between securitizing attempts and the broader public. A good match between an official securitizing narrative and journalistic representations can contribute to acceptance of the call for war in larger sections of society. If the official securitizing narrative is detailed, amplified and repeated in the media, this contributes to the dissemination and discursive prominence of the narrative in the broader public. This is a crucial aspect of how war becomes acceptable, which will be evaluated in this chapter. I turn first to the scrutiny of how Chechnya was represented in journalistic accounts that autumn. Yet again, that is in an effort to uncover the intersubjective nature of the securitizing process and to explain how the new war against Chechnya came to make so much sense in Russia.

As the structure of this chapter indicates, alternative positions on 'Chechnya' were non-existent in journalistic accounts during autumn 1999. The pages of the Russian newspapers studied here did not contain discursive struggles about what meaning to attach to 'Chechnya' broadly speaking – all the articles reviewed fit into the 'discourse of war' in one way or another. Instead, we should ask: *how* did journalistic accounts contribute to make war acceptable?

This chapter investigates how the various parts of the securitizing narrative ('nature of the threat', 'point of no return' and 'way out') are represented in journalistic accounts of the time, as well as how 'Russia' is represented. It also asks whether the 'Chechens' as such are identified as a dangerous group of people. Throughout, journalistic representations are compared with those of the official securitizing narrative to identify similarities and differences. As in the previous chapter on expert language, I will also touch upon the question of what *function* journalistic representations have in relation to official language. How do journalistic accounts substantiate or negate official claims? Do they contribute to the construction of 'Chechnya' on the basis of a particular authority or credibility, or in a particular style? In the concluding section, I sum up and address the question of how successfully the new understanding of Chechnya as an existential terrorist threat was transmitted to the broader Russian public.

The Chechen fighter gets a face

A brief overview of the adjectives most frequently attached to Chechnya/ Chechen fighters shows a fairly clear pattern. We can find more neutral terms such as 'fighters' *(boyeviki)* or 'illegally armed formations', but 'terrorists' or 'the Chechen terrorists', 'Chechen extremists', 'bandits' or 'Chechen bandits' far outnumber these.[1] With the campaign against Chechnya getting underway, media accounts also constantly refer to how military operations are undertaken to 'destroy terrorist bases', or 'liberate' territory from 'the terrorists'. They also repeat the 'fact' that the final aim of the military operation is to destroy all terrorists on the territory of Ichkeriya/Chechnya.[2]

Thus, newspaper accounts served to confirm the official argument that Chechnya was a 'terrorist' threat and to reify such an identification of Chechnya over time. Moreover – and perhaps not surprising given our findings on media discourse in the interwar period (Chapter 4) – newspaper accounts began equating the terrorist threat with Chechnya immediately after the bombings in September: they were less hesitant than official accounts.[3] When the military campaign was well underway and criticism from the West became increasingly vocal, Russian journalists seemed to insist on the representation of Chechnya as an international terrorist threat. According to *NeGa*'s key journalists on the North Caucasus, the past decade had seen the 'total militarization of Chechnya and establishment of a semi-legitimate criminal terrorist regime' which had 'turned the country into a safe haven for international terrorists'.[4]

By presenting details, facts and stories, newspaper reporting from the front serves to give substance and content to more general labels such as 'terrorists' and 'bandits' – or to contradict them. The general impression has been that reporting from the First Chechen War contradicted official claims that Russia was facing a threat in Chechnya, often by giving both Chechen and Russian soldiers a human face and by showing the common grievances of both sides. Now, however, the general impression is that Russian journalists, through the language used in their writings, contributed to giving substance to and expanding on the official claim that Chechnya was an *existential terrorist threat*. Alternative images representing Chechnya as less threatening are difficult to find.

With bombs falling over Chechen territory in September 1999, the pages of *NeGa* and *RoGa* have hardly any reporting on refugees or casualties inside Chechnya. Quite typical is a front-page report in *NeGa* titled 'Is it really true?' The piece reports that MVD (Ministry for Internal Affairs) sources had found a video tape showing how a Russian pilot from a plane downed over Chechnya had been treated by Chechen fighters: 'you hear the cries of the fighters: "Impale him!" and the pilot's reply "No, not that!" The pilot has a sad destiny – one of the bandits slit his throat.' The report concludes that, if the video is not fake, 'the bandits killed him like beasts'.[5] An interview with Chechen fighter Khunkarpashi

Israpilov a few days later, to 'reveal the logics guiding the thinking and actions of the Chechen fighters' leadership', supplements the story of the actions Chechen fighters are capable of. The journalist sums up Israpilov's answers to his questions, noting that

> there is the idea of the "Greater Chechnya", that Moscow mistreats the Dagestanis and totally absurd accusations against Russia that she is forcing Chechens to drink, swear and lie. In a word, a real mix of pretensions, fantasies and outright lies.

The pitch in *NeGa* reporting on Chechnya the first few days after the bomb explosions in Russian cities is thus that not only are the Chechen fighters capable of gruesome deeds, they are also unreliable and even crazy.

This is quite similar to that found on the front pages of *RoGa*. For example, one report titled 'The throat-cutters are making money in hard currency' is followed by a description of how:

> bandit terror is unleashed against those who meet the Russian forces with a sigh of relief. The throat cutters without nationality and faith are killing old people, shooting women and children, executing teachers in front of the pupils – these are facts, taped on film, seen by millions of viewers.[6]

And on the next page under the headline 'Nothing is forgotten':

> in the Cossack *stanitsas* (villages) the bandits carried out ethnic cleansing in accordance with the new Chechen laws. According to official information from the Ministry of National and Regional Affairs, more than 21,000 Russians were executed after the military operations in Chechnya. The bandits stole more than 100,000 flats and houses from Russians, but also from Ingush and Dagestanis. The robbers sent more than 50,000 of their neighbours into slavery. The Ichkerian slaves bent their backs while building roads in the high mountains.... This is still going on in the regions where the federal army is not in control.[7]

On the whole, more general journalistic accounts representing the Chechen fighters as brutal and willing to use 'illegal' weapons (such as gas or nuclear weapons) and tactics (such as abduction, slavery, civilians as human shields, rape, suicide bombing) were widespread in *NeGa* and *RoGa* that autumn.[8] Representations in *NeGa* might have been expected to be less stark and demonizing than in *RoGa*, given that *RoGa* can be considered an official mouthpiece – but that is not always the case.

The construction of the brutal and inhumane Chechen fighter in journalistic accounts is amplified on the pages of Russian newspapers by printing FSB or Ministry of Defence accounts such as the one quoted below in full, without any critical comments by journalists reporting on the North Caucasus:

Dressed as Federal soldiers they will carry out bestial killings, rapes and armed attacks on peaceful citizens. And these mock soldiers will be influenced by alcohol and drugs.... It has been thought out how small traces would be left at the scenes of crimes, signalling that Federal and MVD forces were the perpetrators. The goal of the operation is to make federal forces look guilty of crime, mass killings, lack of discipline and order, and to create a similar image among the populations of Chechnya, Dagestan and Stavropol.[9]

Moreover, in the few individual stories told from inside Chechnya, the representation of Chechen fighters as brutal and inhumane is expounded on in even greater detail. One article gives the story of an orphaned Chechen girl. Her father, together with a group of other Chechen men, had been shot by the fighters because they had refused to dig trenches. According to the girl, there was a fanatic amongst the bandits who were roaming about the village. He liked to amuse himself while intoxicated, walking the streets and poking local passers-by in the stomach with his bayonet. The day before the bombing started, the fighters left the village.[10]

'Brutal', 'inhuman', 'unreliable' and 'treacherous' are key qualities attached to the Chechen fighter in journalistic language, supplemented by the familiar representation of him as a 'criminal' capable of doing anything for money. Under the headline 'Criminals are running the ball', *RoGa* asks: 'how can such a small country fight against one of the biggest military powers in the world? The answer is the combination of a few elements: kidnapping of people, theft of oil and forgery of foreign currency'. The article continues:

> as the forces are being concentrated on both sides of the Terek, the Chechen bandits are again fixing their eyes on the dirty trafficking in hostages. Chechen bandits arm themselves with money from hostage-taking; they demand 5,000 for ordinary Russian people and several millions for foreigners ... Chechens who abduct people make 10 million dollars and possibly more.... The Chechen war is continuously fed by oil, which the Chechens steal ... yet another source of income is pensions and other transfers to Chechnya from Russia to revive the destroyed economy, amounting to 3.5 billion dollars. In the past ten months, Chechen bands were caught with more than 1 million false dollars.[11]

Thus, journalistic accounts in themselves, supplemented by uncritical references to official sources, construct an image of the Chechen fighter that blends the older view of him as a criminal with stark presentations of being inhumane and cruel. This is fairly similar to the qualities given to the Chechen threat in official language. Thus we see that journalistic accounts that autumn served to substantiate, not negate, the official claim that this enemy 'does not have a conscience, shows no sorrow, and is without honour' (as quoted in Chapter 5).

Further, several journalistic accounts make the Chechen threat stand out as even more threatening by offering the combination of 'inhumane' and 'cruel' with its being 'professional', 'well trained' and with 'far-reaching plans', as indicated in official representations. The idea of underlying professionalism and careful planning is bolstered by extensive quotations (without independent journalist comment or evaluation) from official information which revealed the actions and aims of the Chechen fighters. For instance, one account refers to the finding of documents in Dagestan showing how one 'group of deeply conspiratorial people had attended special training in camps on Chechen territory and in Afghanistan.... Their aim – to infiltrate state structures and societal structures'.[12] Other accounts report of Chechen extremists 'planning to establish band formations ... consisting primarily of young people under 18 years ... planning new terrorist acts on Russian territory ... also against nuclear installations ... and by using terrorists with Slavic appearance and women'[13] or 'the extremists are planning to carry out some sharp operations ... Basayev and Khattab are planning a major provocation against Russian forces and the populations of Chechnya, Dagestan and Stavropol'.[14]

In newspaper accounts, as in official language, this representation of the Chechen fighter as inhumane, but professional and well trained, is combined with references to the *magnitude* of the Chechen threat. Even before the Russian leadership had elaborated on this point, *NeGa* journalist Andrey Serenko constructed Chechnya/Wahhabism as a huge threat which might engulf the entire Russian Federation. He noted that 'ever new subjects of the Russian Federation are gradually being drawn into the conflict orbit' and identified 'the existence of tight-knit and influential Caucasian diasporas; the Caucasians strengthening their position in the criminal sphere and the existence of Wahhabi strongholds' as preconditions for such a development, with Volgograd oblast as an example.[15]

The official claim that a 'war' had been declared on Russia was immediately repeated in journalistic accounts. On 15 September, *NeGa* journalist Andrey Kamakin re-phrased and confirmed official representations, writing:

> We have all been declared war upon. Such a war we have not faced before: terrorist war. Yes, there was Budennovsk and Kizlyar, Vladikavkaz and Nalchik, there were ten different big and small terrorist attacks. But all of them were localized in single, separate regions, the rest of Russia slept in peace. Today the war may enter any house, just like it entered two houses in Moscow. The war has no frontiers. The goal of the terrorists is to destabilize the country ... who declared this war? – Everybody knows! Apart from the authors of exotic versions, few doubt that the acts of terror are a continuation of the confrontation in Chechnya and Dagestan.[16]

Similarly, after the explosion in Volgodonsk on 16 September, media accounts linked the terrorist attack to violence in various places throughout the Northern Caucasus, creating an image of a war threatening to engulf the entire Russian state, and squarely stating that 'Chechnya is the centre of terrorism'.[17]

Apart from representations in journalistic accounts, the choice and placement of pictures and headlines further helped to construct the threat as huge and omnipresent. News on terror was constant front-page stuff, and constantly expanded with headlines such as 'The terrorist front on the water'.[18] *NeGa* and *RoGa* accounts of key political events were dominated by references to Russian leadership action against the terrorist threat, for example: 'The main part of their conversation [between the President and Prime Minister] concerned the fight against terrorism'.[19] The bias in such accounts served to underline the gravity of the threat, by suggesting that most political activity *had* to be directed towards fighting the threat.

Journalistic accounts often featured references to a more distant but related enemy, which served to amplify the threat even more. Nor should this be surprising, since the identification of Chechnya with the international terrorist threat was more articulated in media accounts than in official language during the period between the wars.

Journalistic accounts during autumn 1999 posited as 'common sense' a link between the bombings in Moscow, Buynaksk and Volgodonsk to the war in Dagestan via the Chechen threat and further on 'Saudi Arabia', 'Osama bin Laden' or 'Islamic extremism'.[20] Although a few early articles questioned the hypothesis that Bin Laden was the source behind the Chechen threat, instead emphasizing internal Russian problems as the driving force,[21] this did not become a dominant theme on the pages of *NeGa*. Its headlines and articles increasingly represented Chechen fighters and Osama bin Laden as one and the same phenomenon, as shown by the front-page headline as 'Where is Basayev? The whereabouts of 'terrorist no. 1' has not been established, but 'terrorist no. 2' has appeared'.[22] This equation was given content in *NeGa* reporting by frequent references to sources in the FSB. Such accounts would link the 'extremists' in Chechnya to well-funded ideological Wahhabi bases in Azerbaijan and further to 'extremist' organizations in the Arab world and 'well-known terrorist movements like the Muslim Brotherhood'.[23] FSB accounts frequently linked Chechnya to 'Afghanistan', 'the Taliban', 'international terrorism' and 'Osama bin Laden'.[24]

In *RoGa* reporting, this merging of the Chechen and international terrorist threats was even more evident. In particular, there were accounts detailing the linkages between the two and creating the image of Osama bin Laden as the resourceful master and conductor of the Chechen insurgency.[25]

Thus, newspaper discourse during autumn 1999 contributed to constructing the Chechen fighters/threat as a combination of inhuman, cruel but competent, and overwhelming in magnitude, and so closely linked to the world's terrorist no. 1 as to make Chechnya merely an offshoot of this global threat. This is a mix which places the threat representation somewhere close to the top of the scale in terms of danger. These journalistic representations of Chechen fighters undoubtedly add up to a construction of them as constituting an 'existential threat' against Russia. Moreover, the combination of danger and magnitude constructs the situation for Russia as precarious and at a 'point of no return'. Taking up the fight is presented as necessary to secure Russia's future existence. This journalistic representation of both the nature of the threat and of Russia as standing at a point of no return is thus quite similar to official representations during autumn 1999.

Are all Chechens terrorists?

On the other hand, we can also find notable differences between official and journalistic representations of the Chechen threat. While official language usually tried to point out that the 'Chechens' as such should not be considered as part of the threat, journalistic accounts frequently failed to distinguish between Chechen 'bandits' and 'terrorists' and Chechens in general. Often this lack of distinction is explicit, as in expressions like 'Chechens who abduct people make 10 million dollars and possibly more.... The Chechen war is continuously fed by oil, which the Chechens steal',[26] 'Lebed had nearly betrayed Russia with the Chechens',[27] or when 'Chechens' are singled out as a special category of people and associated with crime and violence.[28] At other times, the equation of terrorist or criminal with Chechen is less explicit and more a result of the constant co-occurrence of words like 'Chechnya' and 'Chechen' with 'terror'. Not only individual articles, but also pictures, headlines and the placing of these contribute to the construction of Chechens as dangerous. Nearly every day under the heading 'Terror', accompanied by reports on the evolving war in Chechnya, *NeGa* posted reports on how the war on terror was being pursued across the Russian Federation, with subtexts such as 'Two criminals have been detained who were planning to commit a terrorist act in Vladikavkaz'[29] or 'The terrorists have been put on trial'.[30] 'Chechnya' and 'terror' are inextricably interlinked. How far is the leap from this linkage to the 'Chechen' as the 'terrorist' or the 'criminal'?

In my view, the result of such direct and indirect equations of Chechens with terror and violence over time was that any Chechen was constituted as dangerous. This representation stands out even more strongly because there were very few alternative representations of the 'Chechen' during autumn 1999. One version did seek to distinguish between the civilian population, together with certain Chechen pro-Russian actors, and the dehumanized Chechen terrorists or bandits – but the distinguishing feature of that version of the Chechen was his or her 'Russianness'.[31]

For example, Malik Saydullayev, the Chechen chosen by the Kremlin in autumn 1999 to head the State Council based on the 1996 Zavgayev parliament, was presented as a successful businessman fostered in the common Russian milieu of the 1990s. *NeGa* argued that, given the absence on Chechen territory of 'intellectual resources' capable of creating a 'civilized' society or any kind of 'development', Chechnya can:

> be saved only by the intellectual and business elite of the Chechen community spread throughout Russia. Precisely the Chechen diaspora, which has kept in touch with its historic 'small homeland', and at the same time is linked into the Russian economic structure and the Russian cultural

sphere, can lead their nation, after this nation has rejected the power of the present criminal and militant superstructure.[32]

A similar equation of civilian 'Chechens' with 'Russia' was made in reports from the field as Russian soldiers were advancing into Chechnya. The Chechen civilian population was represented as welcoming the Russian forces as liberators:

> when the Russian army entered the territory of Ichkeriya this autumn, it was met in a totally different manner than in 1994. Not with bayonets, curses, or stones from children's hands, but with a silent hope: has the liberation from the bandit yoke begun, will the forgotten peaceful life return?[33]

One account even stated explicitly: 'not all Chechens are like the leaders of international terrorist bands, who don't care who they kill. Here the Federals might find support amongst those who want to govern Chechnya independently, within Russia'.[34]

Despite this slim acknowledgement of the existence of 'acceptable' Chechens, the heavy construction of the 'suffering Chechen' which had put Chechens on a par with Russians as fellow human beings and had dominated reporting during the First Chechen War was barely present in autumn 1999. There were a few articles that reported casualties and destruction in Chechnya and even interviews with Ichkerian representatives that spoke about the results of the bombardments in the early phase of the war. These accounts contradicted the official Russian claims that only military and technical targets were hit.[35] But even these few accounts offer no details that could substantiate a representation of 'the suffering Chechen'.[36] Moreover, as the war rolled on from October, there were no such inside reports in the pages of *NeGa*.[37] As Chapters 10, 11 and 12 will show, this pattern of reporting continued. Even as potentially 'shocking events' were revealed, such as the bombing of civilian targets or atrocities committed against civilians, words representing the Chechens as victims and Russia as the guilty party did not return to the pages of Russian newspapers.

This silence on Chechen suffering was supplemented by the absence of corresponding visual images. The entire galleries of *NeGa* and *RoGa* photos representing the battlefield in Chechnya during October were shot through the barrel of the Russian federal gun. At the same time, pictures of and stories about the victims and relatives of the terrorist attacks in Moscow and Volgodonsk in September were posted in *NeGa* long into October. These articles reported at length on the sufferings of the victims of the terrorist attacks, naming either 'Chechens' or 'Caucasians' as the perpetrators.[38]

Taken together, the absence of words and pictures representing the Chechens as suffering, fellow human beings, combined with constant linkage of 'Chechen' to violence and crime constructed a one-sidedly negative image. Even if they are not represented in identical fashion, it becomes quite difficult to distinguish 'Chechen' from 'Chechen fighter', and so there are few nuances in the construction of the

Chechen Other. As discussed further in Chapters 10, 11 and 12, this merging of everything Chechen into one category of 'dangerous', if repeated over time and naturalized, makes it possible to undertake certain practices against this group, practices that would not be acceptable against other groups. Thus, we find a logical connection between the construction of Chechens discussed here and the seemingly accepted practice of detaining 'suspicious people on the southern border', when these were Chechens.[39]

A key question, discussed in previous chapters, concerns representations of the Ichkerian President Aslan Maskhadov. To what extent was he merged into the same category as the Chechen fighters in journalistic accounts? In late September, most *NeGa* accounts still presented Maskhadov as an authoritative and reliable person, and indicated negotiation and cooperation with him as the only viable path.[40] *NeGa*'s top journalists on the North Caucasus characterized as 'absurd' Putin's statement of 1 October that the 1996 Chechen parliament was the 'only legal organ of power in Chechnya' (thus discounting the legitimacy of Maskhadov)'.[41]

In the main, however, the Ichkerian leadership was increasingly linked to terms like 'terror' and 'abduction'.[42] And the official claim that Maskhadov had become a consenter to terrorism was mirrored in the view that as long as 'his cooperation to suppress terrorism was questionable' one should look for another partner.[43] Not surprisingly, given its status as a government organ, *RoGa* represented Maskhadov in a fashion even more similar to official representations, stating on the front page: 'Maskhadov has become the obedient toy of incorrigible bandits and foreign masters'.[44] *RoGa* often directly cited high-ranking military officers, allowing these accounts to dominate representations of Maskhadov. For example, on 13 October, Head of the Russian General Staff Anatoly Kvashnin was quoted at length, under the headline 'The double standard of Aslan Maskhadov'.[45] Here Maskhadov was represented as unreliable, criminal and brutal, and his leadership as the source of lawlessness and chaos.

Most striking, however, was how Maskhadov simply disappeared from the pages of *NeGa*. While there were plenty of words demonizing the radical warlords such as Basayev, Khattab and Raduyev and reporting on Chechen warlords who wanted to defect or cooperate with the Federal forces, Maskhadov was left without a distinct face.[46] Instead he was merged into the general representation of the Chechen leadership and its resistance to the military operation as 'terrorist' in accounts such as this: 'To carry out terror acts against the Federal forces and Russian cities the Chechen leaders of the band formations have prepared special groups of fighters'.[47] Or, details were given of links between the Chechen leadership and Islamic extremist actors, which simply served to subsume the Ichkerian leadership under this greater threat.[48] Thus, we find a fair degree of congruence between official and journalistic representations of Maskhadov. Although not directly named a terrorist, he was closely associated with the terrorist threat and could no longer be trusted.

Violent retribution as 'the way out'

In general, the language of journalists, perhaps more than that of any other audience group studied here, failed to distinguish different versions of 'Chechen' with differing degrees of danger attached to them, instead merging them all into one existential 'Chechen threat'. It should therefore come as no surprise that these journalistic accounts suggest very few alternatives to war or some other type of violent retaliation. Ceasefires and negotiations with the Chechen side are presented as failed strategies. The 'Khasavyurt Accord', symbolizing an alternative, non-violent approach of negotiation and compromise, is repeatedly represented as totally unacceptable, even 'disastrous'.[49] Those in Russia who rejected a military solution were occasionally represented as a 'fifth column' in journalistic accounts.[50]

Yet, despite this discounting of peaceful measures and indirect endorsement of war, it is quite difficult to find *explicit* prescriptions for violent retribution against the Chechen threat in journalistic accounts. This might have been different in television reporting, which relies on oral language and often features strong TV personalities entitled to their own opinion. I find no grounds for claiming that the discourse of journalist Mikhail Leontyev during the TV programme 'Odnako' on the popular Channel One (ORT) national network is representative of the language of television journalists, but at least it can serve as a contrast to the reporting in *NeGa*. According to Leontyev, there was only one effective way of dealing with terror:

> against Basayevs, Raduyevs, Khattabs and others, it is necessary to make the earth burn under their feet, their own earth.... It is necessary to create a cordon sanitaire on Chechen territory, where not a single unchecked vehicle is let in. Not a single Russian soldier should be sent into 'liberated Chechen territory' before they [the Chechens] themselves ask for it. Instead, bomber jets should carry out retaliatory strikes, again and again, until the local population understands that there is a connection between the bandits and themselves, the power and the bombs falling over their heads. Until the Chechens themselves come, hand over the crumbs and say 'What can we do to stop this?'... We need to wage war and win. That's all.... Concerning action against the Chechens I say and repeat – they can all go to hell.[51]

Instead of such direct and emotionally charged prescriptions for dealing with the Chechen threat (even before the authorities had taken action), endorsement of the use of force as the best way to deal with Chechnya was expressed in sober terms as necessary and right in *NeGa* reporting. As noted, *NeGa* had immediately linked the terrorist attacks in Russian cities to Chechnya, declaring that 'Chechnya is the centre of terrorism'. Thus, the conclusion in the same article – that 'solving the Chechnya problem with force is necessary, or else the authorities cannot keep control of the country' – is quite logical.[52] A tough, uncompromising approach was also represented as necessary *over time* for dealing with Chechnya:

If a competent and adequately hard power, which is not lenient, is not introduced to punish the criminals and establish order after the Federal forces have done their job, then all efforts and all the blood which has been spilt will have been wasted.[53]

Moreover, as the war unfolded, *NeGa* expressed endorsement of the war and methods of warfare undertaken by the Russian authorities. When it became evident that a new ground offensive against Chechnya was underway, there came some qualified warnings in the Russian press that a new war could become a catastrophe for Russian power. Interestingly, the most thorough piece to present this argument in *NeGa* (5 October) was published anonymously[54] – followed the next day by an article describing how even the Chechen population supported the Russian military campaign and stating: 'nobody doubts the need to destroy the bandit formations'.[55] During the heavy bombardments of Grozny in October, *NeGa* journalists assured their readers that 'the methodical, but not massive, bombardments achieve their goal. Every day a few people die, and unfortunately also some civilians, but it is impossible to avoid casualties totally'.[56]

Even the practice of '*zachistka*' (cleansing operation), which will be discussed further in Chapter 12 and which in human rights reports was associated with torture, killing and disappearances of civilians, is presented as necessary and just. For example, according to *RoGa* journalist Boris Alekseyev:

the only possible way out is to clean the Chechen soil of the terrorist scum and to build a life free of fear, violence and strife on the liberated territory. This is also in practice what is being done on the territory which the federal forces are taking under control.[57]

It is perhaps not surprising to find the practice of *zachistka* justified in a government newspaper. But *NeGa* accounts also represent this as a logical and just way to fight the Chechen threat, albeit with less forceful wording.[58] Over time, the representation of 'Chechnya' as an existential terrorist threat and different versions of violent retribution as the only possible 'way out' were repeated so often and with so little competition from alternative representations that they became naturalized in press accounts.

Russia as a righteous defender

The reconstruction of the Chechen Other as an existential threat demanding violent retribution was accompanied by a reconstruction of Russian identity in general and by new and more positive representations of the Russian defence and security agencies in particular. The suggestion that the Yeltsin regime itself was behind the terrorist attacks and that Putin as head of the FSB had planned them as a pretext for introducing martial law so that the regime could hold on to power, did make it to the pages of Russian newspapers, *Moskovskiy*

Komsomolets in particular. But it never acquired wider resonance, and *NeGa* increasingly referred to it as an unreasonable conspiracy theory constructed to undermine Russian power.[59]

Instead, the story of the First Chechen War was re-written. While the Chechen side was identified as the culprit and responsible for everything that went wrong from 1994 onwards, Russia and the Russian army were stripped of guilt.[60] Chechen/Federal relations were given as a juxtaposition of abuser/benefactor, for instance:

> The Russian Federation has regularly and continuously supplied the 'uncontrolled territory' with free fuel and electricity, practically stopped guarding the border so that the citizens of 'Ichkeriya' could move freely on the territory of the 'metropolitan', enjoying all the same rights as the citizens of Russia, but absolutely none of the duties.[61]

Such general stories were supplemented by individual accounts. Russian generals, all but demonized during the First Chechen War, now acquired status as heroes. For example, General Anatoly Romanov was introduced in the following way: 'The Chechen campaign from 1994 to 1996 created not only misery and traitors, but also real heroes, including among the political military leadership of the federal forces operating in Chechnya. First among these is General Romanov.' The article goes on to tell how he had to take tough decisions during the First Chechen War, such as attacking the village of Samashki; how he tried to persuade the Chechen side to abide by the July 1995 ceasefire agreement; how his efforts were stopped by a 'terrorist act' and finally how this paved the way for the much less favourable Khasavyurt Accord.[62] The Chechen side is represented in only negative terms, whereas the Russian general stands out as the hero. Even 'Samashki', an event that more than any other event had been represented in a way that made Chechens stand out as 'victims' in the Russian media during the First Chechen War, was here portrayed as both necessary and right.[63]

A similar piece was printed concerning the notorious General Shamanov.[64] The article was titled 'The strategy of decisive force: The Russian Caucasian variant', followed by 'General Shamanov did not want to negotiate with the illegally armed formations in 1996 and he does not want to now'.[65] On the First Chechen War, it notes that 'during the first war in Chechnya, Shamanov was injured, but he kept on fighting nevertheless…. He could have contributed to the defeat of the separatists, had it not been for Lebed's peace deal'. Shamanov is presented as 'brilliant', as a 'military talent' who always makes 'heroic efforts to minimize the spilling of blood'. His current military operations in the north of Chechnya are presented as 'particularly effective'. The fact that Shamanov in October 1999 occupied/ liberated Chechen territory far beyond the cordon sanitaire determined by the Russian political leadership was not represented as a problem.

A piece on the Commander of the United Federal Forces Viktor Kazantsev also refers to the Khasavyurt Accord as a failure and to Chechnya in the interwar period as one violent zone, with the actions of the 'extremists' building up to the incursion into Dagestan.[66] Against this background, it presents the Russian response under General Viktor Kazantsev's command as 'well organized and effective', concluding: 'General Kazantsev has, following a long break, refreshed the account of Russian military victories – a victory so necessary for this country, painfully making its way out of the hardships of these troubled times'. Kazantsev was also reported to have offered to give up his monthly salary if Chechen teachers in the 'liberated territory' of Shelkovskogo would return to teaching the children who had not been able to go to school for several years.[67]

This rehabilitation of the previously tainted reputations of these Generals and their new status as heroes was echoed in representations of the Russian soldier. He was now presented as being motivated by the 'pain for those who died as a result of terrorist attacks' and as dutiful and self-sacrificing, contrasted to the Chechen fighters.[68] If Russian soldiers had been represented as victims during the First Chechen War, this time round they were portrayed as winners, 'all of them are convinced that it is necessary to fight to the victorious end'.[69] These linguistic representations were supplemented by symbolic acts constituting Russian soldiers as heroes. As early as 16 October 1999, *NeGa* could report that, since the beginning of the military action in Dagestan, 2318 people had been decorated with 'state honours' and six servicemen had been made 'Heroes of the Russian Federation'. On 20 October, as Russian ground troops were entering Grozny, the paper noted on its front page that Prime Minister Putin had flown a Su-25 fighter jet to North Caucasus to decorate the pilots with 'state honours'. Together with 'warm words for the pilots he (Putin) also thanked the aero-technicians for making possible "minimal losses among the peaceful population"'.[70]

Not only individual soldiers were constituted as heroes in *NeGa* accounts. The entire effort of Russian power in the evolving campaign was generally presented as a civilized undertaking aimed at saving lives. When the ground offensive was underway in Chechnya, media accounts described the campaign as orderly and successful, underlining how the liberated areas of Chechnya received humanitarian help on orders from the Minister of Defence Igor Sergeyev. In contrast to the chaotic and frightening images of the Chechen opponent given in the press, Russian power was 'starting to establish legal state power in the republic ... the refugees were beginning to return home'.[71] The Russian army was described as striving to avoid civilian casualties, with Russian legal authorities guaranteeing that 'every case of illegal action by the servicemen against the civilian population will be followed by a criminal investigation'.[72]

If anything, the chronicles of events in Chechnya in *RoGa* were stronger than those of *NeGa* in representing the Russian army as humanitarian saviours of the Chechen civilian population, while vilifying the actions of the 'terrorists'.[73] Many *NeGa* and *RoGa* accounts seem to be based on information from official sources which inevitably represent Russian power in Chechnya as reliable, orderly, lawful and good.[74]

Finally, journalistic accounts conveyed a sense that 'we are in this together'. Although they did not directly present 'unity' as a key Russian quality in the same way as expert accounts did, they did represent all of Russia as united against the Chechen threat. Forces working in North Caucasus – whether Federal, MVD, FSB or FPS (Federal Border Service) – were described as being supported by and working together with the 'people'. Slogans from the Great Patriotic War such as 'The entire nation is guarding the border' were dusted off.[75]

An interesting aspect, given the recent expansion of violent insurgency from Chechnya into neighbouring Muslim republics, was how Dagestan was represented as part of this Russian unity. 'Dagestan' and 'Dagestanis' were said to 'demonstrate the motivation of those North Caucasian nations who wanted to live inside Russia and in friendship with Russians'.[76] Several journalistic accounts described the people of Dagestan as brave and loyal, and as fellow victims of Chechen lawlessness.[77] Stavropol, another neighbouring republic, was identified similarly as a victim of Chechen violence and as a loyal and brave Russian subject.[78] In this way a clear geographical divide was constructed between 'Russia' and 'Chechnya', placing other federal subjects as unified against Chechnya.[79]

Conclusion

In line with the meta-theoretical perspective which informs this book, I have argued that 'audience acceptance' takes the form of an ongoing process of legitimation to which the putative 'audiences' themselves contribute. The linguistic variations, inventions and re-articulations found across political elite, expert and journalistic texts underscore the intersubjective nature of the process which led to broad agreement on Chechnya as constituting an existential terrorist threat, and on the necessity of a new war. The confirmation of the official narrative in most political elite, expert and journalistic representations during autumn 1999 was a re-articulation of this narrative, which both inserted and rejected certain aspects of the threat as presented in the official language. The official securitizing narrative does not serve as a blueprint which the audience either accepts or rejects. The hegemony of the 'discourse of war' was not forged from above: it grew from the sides and from below.

We have seen how journalistic accounts re-articulated Russian identity in discussing Chechnya during autumn 1999. In re-writing the history of the First Chechen War and the interwar period, accounts in *NeGa* and *RoGa* sought to eradicate ideas of Russian guilt, or of Russia as a lenient and compromising power. On the contrary: Russia was now characterized by decisiveness, efficiency and bravery, combined with benevolence and humanitarianism. This reconstruction was undertaken first and foremost through representations of Russian security personnel as concrete expressions of 'Russian power'. Newspaper accounts also gave content to the claim that Russia was *united* against Chechnya in this war – not only by downplaying any discord amongst the various power agencies operating in Chechnya, but also by representing the federation republics close to Chechnya as fellow victims of Chechen violence.

The massive re-invention of those carrying out Russian policy in Chechnya as righteous defenders and benefactors in journalistic accounts spills over into the general re-construction of Russia that autumn. Where expert accounts constructed Russia's moral superiority by use of historical references, journalistic accounts did so by reporting and detailing Russian deeds as the military operation unfolded. This served to underpin and substantiate the 1999 official articulation of Russian identity, projecting Russia as strong, innocent and capable of establishing order. Thus, the official calls for 'Russian unity and strength', a slogan that was to permeate Putin's presidencies in the years ahead, did not ring out as a single voice: this was much more of a collective call.

Similarly, journalistic accounts through stories revealing the atrocious actions committed by Chechen fighters served to substantiate official claims about the brutal and inhuman nature of the threat. Even 'Maskhadov' gradually became identified more with the terrorist Other than with the Russian Self. Journalistic accounts gave credibility to the official securitizing narrative by reporting 'facts' from the ground. By uncovering the specific, wide-ranging plans and mapping the extensive geographical presence of the threat, newspaper accounts made the magnitude of the Chechen threat as part of the international terrorist threat stand out as 'real', no longer just words from the mouth of the Prime Minister.

We have also seen how journalistic accounts played a special role in giving Chechens as such an identity as different and dangerous, far beyond what was indicated in the official narrative: first, because of the many direct and indirect equations of Chechens with terror and violence; second, because of the near-total absence of newspaper reports on the casualties and destruction that would have carried alternative representations of Chechens as human and suffering. This is a key difference between journalistic reporting during the First Chechen War and the Second Chechen War[80] and is particularly important when discussing how journalistic accounts provide a link between official claims and their acceptance among the broader public in times of war. When newspaper accounts merge everything Chechen into the 'existential terrorist threat', without describing the suffering and misery of the Chechen people, they remove one of the most potent mechanisms available for mobilizing a population against war: feelings of identification and compassion with the target.

On the whole, we can see an important difference between the journalistic texts analysed in this chapter and those of the political elite and experts. Whereas alternative positions – in the form of the 'discourse of reconciliation' or of a hybrid position – could be identified in political elite and expert texts, this was not the case with journalistic accounts. Such alternative positions may well have existed in more marginal newspapers, but they did not find expression in the editions of *NeGa* and *RoGa* analysed here. I do not see this bias as solely the result of tighter media control. Restrictions on the media were introduced gradually and were not in full force during autumn 1999. The congruence between media discourse and official discourse in terms of representing the Chechen threat as an existential threat necessitating a policy of violent retribution was produced in a fairly open field.

Nevertheless, the most striking finding in the majority of all texts reviewed in this book is that they projected a sharp juxtaposition between 'Chechnya' and everything Chechen as an overwhelming, existential threat to a righteous, superior and good 'Russia'. They transmitted a sense of Russia as at *a point of no return*, as in an emergency situation in which radical emergency measures were the only logical *way out*. While most texts reviewed in this book convey this sense of urgency through the nouns, verbs and adjectives that are attached to Chechnya and Russia, newspaper accounts also contributed to this sense of urgency by the placement of pictures and the use of headlines.

Spreading the word

The words of politicians, experts and journalists investigated in this book played a key role in transmitting a new core understanding of 'Chechnya' and 'Russia' to other audiences beyond themselves. In this way they provided an important link between official claims and the acceptance of such claims by the broader public in times of war. With the near clear-cut dichotomy created by merging everything Chechen into one category of 'dangerous' on the one hand, and with a righteous and benevolent 'Russia' united against this threat on the other, war must have appeared both logical and acceptable for those who related to it through the words of politicians, experts and journalists. Against this background, the public opinion polls indicating strong support in autumn 1999 for undertaking violent measures against Chechnya indeed become understandable.

Taken together, opinion polls that autumn indicated firm endorsement of the security claims made by the Russian leadership. Polls conducted between 17 and 21 September following the terrorist attacks showed that most Russians interviewed were convinced that the attacks were committed by Chechen fighters/Wahhabis; that over 80 per cent feared falling victim to such attacks; and that there was widespread acceptance for measures that must be considered, in the terminology of securitization theory, as extraordinary and beyond the boundaries of conventional rules. Moreover, 75 per cent agreed with the statement that 'accounts of Chechen firms should be frozen, searches of their offices and storage facilities carried out', and 63.7 per cent agreed with the statement 'all Chechens should be expelled from Russia to Chechnya'. At this time, more than 64 per cent agreed that Chechnya should be given a choice: 'stop the terrorist acts or face massive bombardments of the Republic's territory'.[81] A few weeks later, 50 per cent of those surveyed placed all the guilt for the terrorist acts on 'Chechen band formations', 20 per cent on 'centres of international terrorism', 18 per cent on 'representatives of Russian oligarchs' and 11% per cent on 'Russian special services'.[82] If figures in opinion polls are used as a measure of audience acceptance, we can say there was acceptance of the official securitizing narrative as the second all-out war in Chechnya became reality. Polls conducted by the Russian Public Opinion Foundation (FOM) in November 1999 showed that 64 per cent approved of Russian military actions in

Chechnya, while only 23 per cent disapproved. These figures remained fairly consistent through to the beginning of 2001.[83]

Thus, we find fairly broad agreement across Russian society on Chechnya as posing an existential threat to Russia. In Chapters 11 and 12, this finding is reconfirmed when we investigate representations of Chechnya and Russia among police and security personnel. The 1999 war against Chechnya seems to have become an *acceptable undertaking* in the eyes of most Russians. However, according to the perspective informing this book, this type of consensus or acceptance by the audience can never be considered a *stable* arrangement. When public legitimation is the result of a transactional process in which both speaker (in this case the Russian leadership) and audience take part, it can also unravel via another transactional process. Securitizing claims must be continually reproduced: no object can be so firmly established as an existential threat necessitating extra political action that it cannot be challenged.

The previous three chapters have shown, however, how the 'discourse of war' permeated and dominated linguistic expressions in Russian society during autumn 1999. The key argument is that this discursive hegemony made the undertaking of violent retribution against Chechnya possible. The physical practices of war investigated in the next three chapters would hardly have been tolerated or accepted were it not for the extensive and many-layered re-phrasing of Chechen and Russian identity inherent in linguistic practices during autumn 1999.

Notes

1 'Psikhologicheskaya voyna v Chechne razgorayetsya', *NeGa*, 26 October 1999; 'Putin predlagayet novyy plan Chechenskogo uregulirovaniya', *NeGa*, 15 September 1999; 'Buynaksk, dva raza Moskva, teper' Volgodonsk: gde dal'she?', *NeGa*, 17 September 1999; 'Zadacha federal'noy vlasti...', *RoGa*, 12 October 1999; 'Kuda teper' napravyatsya boyeviki', *NeGa*, 21 September 1999; 'Rossiya snimayet ekonomicheskuyu blokadu s Abkhazii', *NeGa*, 22 September 1999; 'Plany suchoputnoy operatsii', *NeGa*, 28 September 1999; 'Gde zhe Basayev?' *NeGa*, 13 October 1999; 'Zakon na storone Federalov', *NeGa*, 13 October 1999; 'Novaya taktika Moskvy', *NeGa*, 15 October 1999; 'V Chechne nachat vtoroy etap voyskovoy operatsii', *NeGa*, 19 October 1999; 'Samyye ozhestochennyye boi yeshche vperedi', *NeGa*, 21 October 1999. Articles in *RoGa* used 'terrorist' 'bandit' and 'extremist' even more often than *NeGa* ('Voyska na Tereke. Chto dal'she?', *RoGa*, 13 October 1999; 'Goryachaya Khronika', *RoGa*, 15 October 1999).

2 'Voyna v Chechne vozmozhna', *NeGa*, 24 September 1999; 'V Chechne hachat vtoroy etap voyskovoy operatsii', *NeGa*, 19 October 1999; 'Samoye trudnoye – vperedi', *NeGa*, 27 October 1999; 'Zadacha federal'noy vlasti...', *RoGa* 12 October 1999.

3 'Kavkazskiye plenniki', *NeGa*, 14 September 1999; 'Putin predlagayet novyy plan Chechenskogo uregulirovaniya', *NeGa*, 15 September 1999.

4 Ilya Maksakov and Zagid Varisov 'Al'ternativy na primeneniye sily net'; *NeGa*, 27 November 1999.

5 'Neuzheli eto Pravda?' front page, *NeGa*, 14 October 1999.

6 'Golovorezy otrabatyvayut valyutnuyu zarplatu', *RoGa*, 19 October 1999.

7 'Nichto ne zabyto', *RoGa*, 19 October 1999.

8 'Blokadnyy Grozny', *NeGa*, 29 October 1999; 'Federal'nyye voyska podoshli k Groznomu', *NeGa*, 21 October 1999; 'Psikhologicheskaya voyna v Chechne razgorayetsya', *NeGa*, 26 October 1999; 'Vokryg Groznovo szhimayetsya koltso', *NeGa*, 28 October 1999; 'Gde zhe Basayev?', *NeGa*, 13 October 1999; 'Voyska na Tereke. Chto dal'she?', *RoGa*, 13 October 1999.

9 'Plan Rossiyskikh voyennykh – 'upolovinit'' Chechnyu', *NeGa*, 29 September 1999. See also 'Samoye trudnoye – vperedi', *NeGa*, 27 October 1999. 'Federal'nyye voyska podoshli k Groznomu', *NeGa*, 21 October 1999 is another example. This tendency to cite official information directly and without comment was even more striking in *RoGa* articles (eg. 'Kavkazskiy uzel: 18 October', *RoGa*, 19 October 1999).

10 'Federal'nyye voyska podoshli k Groznomu', *NeGa*, 21 October 1999.

11 'Kriminal pravit bal', *RoGa*, 16 October 1999. Similar representations on stealing, hostage-taking and slavery are found in 'Problema terrorizma v Rossii tesno svyazana s nelegal'noy migratsiey', *NeGa*, 2 October 1999 and 'Psikhologicheskaya voyna v Chechne razgorayetsya', *NeGa*, 26 October 1999.

12 'Dvoynaya opasnost' ', *NeGa*, 18 September 1999.

13 'Gde zhe Basayev?', *NeGa*, 13 October 1999. Reports on how the 'bandits' were planning to use tactics such as suicide bombing, nuclear weapons, rapes and abductions often against innocent people were frequent. (See for example 'Voyska na Tereke. Chto dal'she?', *RoGa*, 13 October 1999).

14 'Plan Rossiyskikh voyennykh – 'upolovinit'' Chechnyu', *NeGa*, 29 September 1999. Similar reports citing FSB information in 'Samoye trudnoye – vperedi', *NeGa*, 27 October 1999 and 'Pushechnoye myaso dlya religioznykh voyn gotovyat v Azerbaydzhane', *NeGa*, 7 October 1999. This tendency to merely cite official information was even more striking in *RoGa* articles (e.g.'Kavkazskiy uzel: 18 October' 19 October 1999.)

15 'Kavkazskiye plenniki', *NeGa*, 14 September 1999.

16 'Putin predlagayet novyy plan Chechenskogo uregulirovaniya', *NeGa*, 15 September, 1999.

17 'Buynaksk, dva raza Moskva, teper' Volgodonsk: gde dal'she?', *NeGa*, 17 September 1999.

18 'Terroristicheski front na vode', *NeGa*, 8 October 1999.

19 'Vo-p'ervykh, bor'ba s terrorom', *RoGa*, 19 October 1999.

20 In the days following the bomb explosion in Moscow on 13 September, for example, several media articles linked the bomb explosions to Chechnya and to Osama bin Laden and the 1998 terror attacks against U.S. embassies in Kenya and Tanzania, citing Western media outlets ('Mir eshchë ne stalkivalsya s takoy zhestokost'yu terroristov', *NeGa*, 14 September 1999). See also 'Saudovskiy sled v Dagestanskom konflikte', *NeGa*, 18 September 1999.

21 'Kto stoit za Separatistami', *NeGa*, 15 September 1999.

22 'Gde zhe Basayev?', *NeGa*, 13 October 1999. See also 'Psikhologicheskaya voyna v Chechne razgorayetsya', *NeGa*, 26 October 1999 and 'Snachala – chistyy islam, zatem – gryaznaya voyna', *NG Regiony*, 28 September 1999.

23 'Pushechnoye myaso dlya religioznykh voyn gotovyat v Azerbaydzhane', *NeGa*, 7 October 1999.

24 'Voyna za svoy schët', *NeGa*, 20 October 1999 and 'Vokrug Groznogo szhimayetsya kol'tso', *NeGa*, 28 October 1999.

25 See for example 'Glavnyy sponsor terroristov', *RoGa*, 15 October 1999.

26 'Kriminal pravit bal', *RoGa*, 16 October 1999. Other examples can be found in 'Novaya taktika Moskvy', *NeGa*, 15 October 1999 and 'Grozny bydut brat' po chastyam', *NeGa*, 22 October 1999.

27 'V Prigranichnykh s Chechney rayonakh rastet chislo bezhentsev', *NeGa*, 6 October 1999.

28 See for example 'Inostrannaya podpitka Ichkerii', *NeGa*, 30 October 1999.

29 'Zaderzhany dva prestupnika', *NeGa*, 6 October 1999.
30 'Terroristov otpravili na skam'io podsudimykh', *NeGa*, 14 October 1999.
31 In this discourse the Chechen nation was represented as an 'indivisible part of the Russian poly-ethnic community. The specificities of Chechen mentality constitute one of the most common archetypes of the all-Russian worldview, it does not lie outside of the broad common standard.' ('Na perekrestkakh Chechenskoy sud'by', *NeGa*, 22 October 1999).
32 'Na perekrestkakh Chechenskoy sud'by', *NeGa*, 22 October 1999.
33 'Golovorezy otrabatyvayut valyutnuyu zarplatu', *RoGa*, 19 October 1999.
34 'Psikhologicheskaya voyna v Chechne razgorayetsya', *NeGa*, 26 October 1999.
35 'V Kreml' cherez Chechnyu?', *Segodnya*, 28 September 1999.
36 'Federal'nyye voyska podoshli k Groznomu', *NeGa*, 21 October 1999 and 'Blokadnyy Grozny', *NeGa*, 29 October 1999.
37 Apart from a small item from Agence France Press on ten dead civilians when the Presidential palace in Grozny was bombed on 21 October (*NeGa*, 22 October 1999).
38 'Ekho vzryva na tikhom Donu', *NeGa*, 26 October 1999.
39 'Podozritel'nykh lovyat na yuzhnoy granitse', *NeGa*, 22 October 1999.
40 'K novoy voyne v Chechne', *NeGa*, 23 September 1999 and 'Voyna bez vykhodnykh', *NeGa*, 25 September 1999 and 'Plan Rossiyskikh voyennykh – 'upolovinit'' Chechnyu', *NeGa*, 29 September 1999.
41 'Nevernyy shag v pravil'nom napravlenii', *NeGa*, 2 October 1999; also 'Taynye i yavnye manevry Moskvy', *NeGa*, 5 October 1999.
42 'Problema terrorizma v Rossii tesno svyazana s nelegal'noy migratsiey', *NeGa*, 2 October 1999.
43 'Kavkazskiye plenniki', *NeGa*, 14 September 1999.
44 'Maskhadov stal poclushnoy igrushkoy v rukakh otpetykh banditov i inostrannykh khozayyev', *RoGa*, 14 October 1999.
45 'Dvoynaya moral' Aslana Maskhadova', *RoGa*, 13 October 1999.
46 This is evident in 'Stanut li boyeviki pomogat' novoy vlasti?', *NeGa*, 20 October 1999 as well as in 'Zadacha federal'noy vlasti', *RoGa*, 12 October 1999.
47 'Strategiya reshitel'noy sily: Rossiysko Kavkazskiy variant', *NeGa*, 22 October 1999. There were exceptions, however. An article by Abdulkhamid Khatuyev ('Blokadnyy Grozny') in *NeGa* on 29 October even used the title 'President of Chechnya' to present Maskhadov.
48 See for example 'Pushechnoye myaso dlya religioznykh voyn gotovyat v Azerbaydzhane', *NeGa*, 7 October 1999.
49 'Strategiya reshitel'noy sily: Rossiysko Kavkazskiy variant', *NeGa*, 22 October 1999. See also, for example, 'V Prigranichnykh s Chechney rayonakh rastet chislo bezhentsev', *NeGa*, 6 October 1999 and 'Yego imya oznachayet pobeditel' ', *NeGa*, 24 September 1999.
50 'Psikhologicheskaya voyna v Chechne razgorayetsya', *NeGa*, 26 October 1999.
51 Cited in 'Chechnya ugrozhayet televedushchemy Leontyevu', *NeGa*, 17 September 1999.
52 'Buynaksk, dva raza Moskva, teper' Volgodonsk: gde dal'she?', *NeGa*, 17 September 1999.
53 'Samyye ozhestochennyye boi yeshche vperedi', *NeGa*, 21 October 1999.
54 'Pobedy nastoyashchiye i mnimyye', *NeGa*, 5 October 1999.
55 'V Prigranichnykh s Chechney rayonakh rastet chislo bezhentsev', *NeGa*, 6 October 1999.
56 'Grozny budut brat' po chastyam', *NeGa*, 22 October 1999.
57 'Zadacha federal'noy vlasti', *RoGa*, 12 October 1999.
58 'V Chechne nachat vtoroy etap voyskovoy operatsii', *NeGa*, 19 October 1999.
59 'Psikhologicheskaya voyna v Chechne razgorayetsya', *NeGa*, 26 October 1999.
60 'V Prigranichnykh s Chechney rayonakh rastet chislo bezhentsev', *NeGa*, 6 October

1999 and 'Strategiya reshitel'noy sily: Rossiysko Kavkazskiy variant', *NeGa*, 22 October 1999.

61 'Samyye ozhestochennyye boi yeshche vperedi', *NeGa*, 21 October 1999.
62 'Geroy proklyatoy voyny', *NeGa*, 8 October 1999.
63 On Samashki see http://en.wikipedia.org/wiki/Samashki_massacre, accessed 17 March 2016.
64 On Shamanov see http://en.wikipedia.org/wiki/Vladimir_Shamanov, accessed 17 March 2016.
65 'Strategiya reshitel'noy sily: Rossiysko Kavkazskiy variant', *NeGa*, 22 October 1999.
66 'Yego imya oznachaet pobeditel'', *NeGa*, 24 September 1999.
67 'V shkoly pod bombami', *RoGa*, 15 October 1999.
68 'Iz Voronezha – v Dagestan', *NeGa*, 23 September 1999, see also 'Voyna za svoy schet', *NeGa*, 20 October 1999.
69 'Federal'nyye voyska podoshli k Groznomu', *NeGa*, 21 October 1999.
70 Ibid.
71 'Osvobozhdena tret' Chechni', *NeGa*, 16 October 1999. Similar accounts were given in 'Zadacha federal'noy vlasti', *RoGa*, 12 October 1999 and 'Gumanitarnoy katastrofy v Ingushetii poka net', *NeGa*, 1 October 1999.
72 'Grozny budut brat' po chastyam', *NeGa*, 22 October 1999.
73 'Chechenskaya khronika, 13 October', *RoGa*, 14 October 1999 and 'Khronika', *RoGa*, 15 October 1999.
74 'Voyna za svoy schet', *NeGa*, 20 October 1999, 'Federal'nyye voyska podoshli k Groznomu', *NeGa*, 21 October 1999 and 'Blokadnyy Grozny', *NeGa*, 29 October 1999.
75 'V Dagestane po prezhnemu nespokoyno', *NeGa*, 29 October 1999; also 'V shkoly pod bombami', *RoGa*, 15 October 1999.
76 'Dvoynaya opasnost'', *NeGa*, 18 September 1999.
77 'Aul v peskakh', *NeGa*, 8 October 1999 and 'Na pomoshch' Dagestanu', *RoGa*, 16 October 1999. The film *A Dagestani Response* was produced by the television and radio company Mir and was widely distributed. It purported to give a picture of the incursion into Dagestan and its aftermath. In the film the Dagestanis were said to be 'ready to fight the enemy with their bare hands … they are even ready to help the Russian forces on Chechen soil'. According to the director, the point of the film was to show how 'the inhabitants of Dagestan, in a national movement of resistance, with weapons in their hands stood up to defend their fatherland'. The director described the Dagestanis who resisted the 'Chechen attack' as 'heroes', while the attackers were consistently referred to as 'Chechens' or 'bandits'. According to the film, rich Akkintsy Chechens were said to have known of the attack before it happened and took care of themselves by sending their families away. In contrast, many Dagestani refugees agreed to the destruction of their villages if that was what it would take to 'annihilate the bandits and establish peace and order in Dagestan' ('Krovavyye s''yemki', *NeGa*, 16 October 1999).
78 'Stavropol ne brosit na proizvol sud'by Chechenskikh sosedey', *RoGa*, 21 October 1999.
79 An exception is Ingushetiya, which had an unclear status in newspaper reporting. For example, the Ingush President was accused of exaggerating the number of refugees from Chechnya.
80 On the changing media situation in Russia, see E. Mikiewicz. 1997. *Changing Channels: Television and the struggle for power in Russia.* New York: Oxford University Press; E. Mikiewicz, 2008. *Television, Power, and the Public in Russia.* Cambridge: Cambridge University Press; S. White and I. McAllister. 2006. Politics and the media in post-Communist Russia. In: *Mass Media and Political Communication in New Democracies*, edited by K. Voltmer. New York: Routledge, 210–227.
81 Курьер 1999 9 Время проведения: 17.09.1999 – 21.09.1999 Число опрошенных:

1545 Вопросов в исследовании: 189 available at http://sofist.socpol.ru/oprview. shtml?en=0 Kur'yer 1999 9 Vremya provedeniya: 17.09.1999 – 21.09.1999 Chislo oproshennykh: 1545 Voprosov v issledovanii: 189 available at http://sofist.socpol.ru/ oprview.shtml?en=0, accessed 15 January 2014.

82   The figures are from a public opinion poll conducted 2–4 October 1999 by the Russian Independent Institute for Social and National Problems, cited in *NeGa*, 14 October 1999, 8.

83   These were the results of nationwide surveys, sample size of 1500 respondents, conducted in 100 localities in 44 regions, territories and republics. The question was: 'Do you approve of Russian military actions in Chechnya, or not?' Available at http://bd. english.fom.ru/report/map/projects/dominant/dominant2002/239_3617/662_3631/209 3_3645/ed020708, and accessed 5 January 2013. Similar figures were found by other prominent polling agencies such as VTsIOM and ROMIR. Whereas in January 1995 54.8 per cent of the population opposed the use of military means in Chechnya and this mood was confirmed in January 1997 by strong support (67 per cent) for the Khasavyurt Peace Agreement, in November 1999, 52 per cent were in favour of establishing constitutional order in Chechnya by use of the army (B.K. Levashov. 2001. *Rossiyskoye obshchestvo i radikal'nye reformy*. Moscow: Akademia, Russian Academy of Science, Institute of Social-political Research, 850–852).

10 Sealing off Chechnya

The broad intersubjective process which evolved during autumn 1999 and brought Chechnya into being as an existential terrorist threat was not limited to the myriad of Russian statements and images reviewed in previous chapters. It also included a whole row of material practices of war undertaken against Chechnya. The next three chapters will lay out what these 'emergency measures', in the terminology of securitization theory, amounted to. Such an account cannot summarize every single Russian policy or practice toward Chechnya from autumn 1999 onward: it must, in line with securitization theory, focus on those that go beyond the 'rules that otherwise have to be obeyed'.[1]

Here we want to identify ways of dealing with Chechnya and practices undertaken against Chechnya and Chechens that would usually have been considered illegitimate, but which seemed called for by this urgent situation. We expect to find two distinct aspects to this moving beyond the rules: first, that measures that had been socially unacceptable only a while ago were suddenly accepted as reasonable and even necessary by the Russian political elite and the Russian public (social rules); second, that measures contrary to the legal foundations of the Russian state or Russian laws became accepted and even explicitly endorsed by the Russian political elite (legal rules). In sum, the next few chapters examine different types of emergency measures that were made possible and legitimate through the representation of Chechnya/Chechens as an existential threat, measures that went beyond the rules that otherwise have to be obeyed.

Given the theory framework of this study, 'emergency measures' are seen as equivalent to 'knowledgeable practices' in post-structuralist discourse theory: They are the material expressions of significative practices, and are seen as complementing these. Thus, 'emergency measures' should be studied by exploring the link between two aspects: the linguistic representations in the securitizing narrative investigated in the previous empirical chapters; and implementation of these in policies and security practices aimed at countering the threat – which is the focus of the next three chapters. My choice of incorporating quotes into the account of material practices below is based on this conceptualization. While linguistic practices have been presented apart from the material practices detailed in these chapters, they are theorized as being intertwined: not because linguistic practices *cause* certain policies or material practices, but because they open up or constrain the range of policies and material practices deemed possible and legitimate. Simultaneously, the material practices are central to the constitution, production and maintenance of the linguistic identity construction that they enact. We will therefore also explore how language (on the micro- and macro-levels) enables and legitimizes material security practices *as they are carried out*. Another key exercise is to reveal how the undertaking of these practices transmits and cements the dominant discourse on Chechnya and Chechens to the micro-levels of Russian society.

The present chapter starts by briefly discussing the immediate endorsement by the Russian Federal Assembly of policies and practices 'beyond rules that otherwise have to be obeyed' explicitly indicated by the Russian leadership at the beginning of the Second Chechen War. It then moves on to investigate the practice of 'sealing off' Chechnya and Chechens from Russia. The chapter considers the physical isolation of the republic and the militarization of the bordering regions, as well as the re-assigning of all relations with Chechnya to the sphere of security. It also discusses how requirements of re-registration for Russian citizens and the fabrication of criminal cases became practices that served to seal Chechens off from Russian cities, constituting them as 'different' and 'dangerous' within Russian society.

Initial endorsement

From the representations of the threat, the 'point of no return' and the 'way out' outlined in the official securitizing narrative (Chapter 5), the radical and concrete emergency action undertaken by the Putin government against Chechnya during 1999 seemed both logical and legitimate. The broad acceptance which the official securitizing narrative enjoyed among the Russian political elite in the Federal Assembly swiftly translated into formal endorsement of new policies and emergency measures against Chechnya when requested, and broad moral endorsement where formalities were deemed unnecessary.[2]

The new plan on Chechnya that Putin presented to the Duma and the Federation Council in September would have been totally unacceptable only months before. It entailed: an 'objective reassessment' of the 1996 Khasavyurt Accord; the imposition of a strict cordon sanitaire along Chechnya's borders; the employment of preventive strikes to 'destroy' all guerrilla bands on Chechen territory; the presentation of an ultimatum to Chechen authorities demanding the extradition of fighters present on Chechen territory; the imposition of a 'special economic regime' in relations with Chechnya; and eventually the creation of a Chechen government in exile.[3]

This would certainly involve moving 'beyond rules that otherwise have to be obeyed' in order to fight off the Chechen threat, in both social and legal terms. In practice, the 'reassessment' of the 1996 Accord meant scrapping it altogether. This political agreement, which epitomized and codified the non-violent relations between Russia and Chechnya, had enjoyed strong support among the Russian audience.[4] Despite the well-known dislike for this peace agreement in the Russian Army, any suggestion of annulling it would not have found broad acceptance prior to summer 1999. The idea of a cordon sanitaire had been proposed early that year by Stepashin, but was dismissed at the time because the deployment of border troops to patrol administrative borders contravened the Russian Constitution. Similarly, the employment of preventive strikes against Chechen territory had been a totally unacceptable measure only one year earlier (as discussed in Chapter 4). Also the final point of the plan, 'the creation of a Chechen government in exile', would have been unacceptable for most in the Russian political elite before summer 1999, because Maskhadov's status as legitimate leader of Chechnya had been indisputable (see Chapter 4).[5]

Now, however, there was support for the government's plan for handling Chechnya and the fight against terrorism in all Russian branches of power. The senators, whose formal support was necessary for the initial use of force against Chechnya, expressed their full support for all the measures proposed.[6] Although the document presented to the Federation Council did not specifically mention the cordon sanitaire, the use of preventive strikes or the imposition of a 'special economic regime', in principle it sanctioned the government's action plan on Chechnya.[7] The Duma (which only the day before had been divided on whether to condemn the incumbent Yeltsin regime) fully endorsed the plan. The press reported that the refrain repeated throughout the session was 'we support you, whatever laws are needed, we will pass them'.[8]

One such measure proposed by the government and endorsed by the Duma was the decision to 'implement all necessary measures to prevent any appearances in the press of representatives of armed formations, war propaganda, calls to encroach on the territorial integrity of the Russian Federation and instigation of social unrest'. Media outlets that did not meet these demands would have their licences revoked.[9] The introduction of this measure was in effect made possible by the common representation of the Chechen adversary as inhuman, extremely dangerous and not entitled to 'a face'. The important point here is how easily such a measure – one which was in contravention of the Constitution as well as the media law – was sanctioned and approved by Russian legislators in autumn 1999.

As important as the endorsement of certain measures was the fact that certain other measures were *not* suggested by the government, and that this state of 'non-measures' was endorsed by the Federation Council and the Duma. Prime Minister Putin's statement that there was no need to introduce a state of emergency is a good example.[10] He urged the politicians in the Duma not to talk about a lack of the necessary legal basis for conducting the struggle against terrorism, and argued that the 1998 law 'On Combating Terrorism' provided a sufficient legal foundation for pursuing the struggle in the Northern Caucasus and in Russia as a whole.[11] And no one in the Duma or in the Federation Council raised the issue. Quite the contrary: many had expressed fears that the Yeltsin regime would introduce a state of emergency as a pretext for postponing elections.

Thus, the definition of the use of force against Chechnya as a 'counterterrorist operation' was accepted and confirmed by Presidential Decree no 1155 of 27 September 1999.[12] No state of emergency or martial law was ever introduced. Presidential Decree no 1155 ordered the government to prepare a resolution that would stipulate the legal foundations of the operation and determine the social guarantees of the servicemen. That was, however, never done. The scare was so intense and the dynamics so swift during autumn 1999 that formalities could be skipped without anyone apparently noticing. Thus, the full ground offensive which was launched against Chechnya on 1 October was not even explicitly proposed by the Russian leadership.[13] No formal endorsement was sought for this part of the new offensive, and the Russian Federal Assembly undertook no formal moves to oppose it.

As to what kinds of forces could legally be used to fight the Chechen threat, the Russian Chief Military Prosecutor held that, whereas the conflict of 1994 had been an internal conflict between *rossiyane* ('Russians' in the non-ethnic sense) and Chechens, and therefore according to the military doctrine called primarily for the use of internal forces, the current conflict was one against 'bands of international terrorists'. Further: 'the terrorists are not only well armed but very well armed. Police forces, also the Ministry for Internal Affairs forces, cannot cope with such bands. The army should fight and destroy them'. Since the goal of the 'international terrorists' when entering Dagestan was

> to break away Dagestan from the Russian Federation ... and since we are faced by hired mercenaries from foreign countries.... In this situation, to defend the territorial integrity of the state, one of the foundations of the Russian constitutional order, the use of Federal Forces, is not only legal but necessary'.[14]

When framed in line with the securitizing narrative, such an operation – including the use of Federal Forces on Russian territory against Russian citizens – was no longer considered unconstitutional.

In sum, then, we find an initial endorsement by the Russian Federal Assembly of emergency measures proposed by the Russian leadership that went 'beyond rules that otherwise have to be obeyed' in social as well as legal terms. The endorsement of the state of non-measures which meant that there were few legal rules to guide conduct in the military/security action ahead was a particularly problematic starting point. How was an anti-terror operation to be conducted, with thousands of troops on thousands of square kilometres against hundreds of fighters, if the rules regulating the conduct of servicemen in a 'state of emergency' or in a 'war' did not apply? This uncertainty on the rules of the game during the Second Chechen War made the parameters for legitimate action drawn up in the securitizing narrative particularly relevant, as we shall see below.

The physical isolation of the republic

During autumn 1999, policies and practices that had seemed beyond the rules half a year earlier were not only formally endorsed, they were also enacted. Imposing a strict cordon sanitaire around Chechnya appeared logical and legitimate, given the new official representations that constructed Chechnya as an extreme, inhuman, well-planned and well-connected danger threatening Russia.

Newspapers ran pictures of huge ditches being carved out around the republic and lined with barbed wire. In the course of a short time, from mid-September, there emerged a near-total militarization of Russian territory bordering on Chechnya. A battalion from the Marine Infantry of the Black Sea Fleet was moved to the Dagestan–Chechen border, and military divisions were dispatched from the Moscow and St Petersburg military oblasts, as well as storm troops, numbering in total 2,500. Their task: 'to destroy bandit and terrorist bases in Chechnya'.[15] Police OMON *(otryad militsii osobogo naznacheniya)* troops were sent to the region from all over the Russian Federation.[16] Eventually three tiers of forces were established to surround Chechnya. The first consisted of Ministry of Interior forces, OMON and police forces. Their task was to 'conduct a hard and systematic control of everybody and everything that crosses the border either way'. The second and third tiers were made up of Ministry of Defence troops, whose task was to 'prevent the movement of band formations and to support the Ministry of Interior forces with firepower if necessary'.[17]

Regional authorities also contributed to this militarization. In the neighbouring *kray* of Stavropol, the Stavropol Security Council adopted a resolution on 17 September not to 'allow bandit incursions from the Chechen side'. Staff centres were established in every region and city of Stavropol, instructed to follow the situation operatively and respond immediately to the situation; all strategic buildings were put under military/security protection; and administrative leaders were instructed to call upon self-defence units to protect the civilian population. More than 3,000 Cossack troops were prepared to secure Stavropol.[18]

The FPS (Federal Border Service) was strengthened considerably (in number of troops and posts) along all federal borders, especially those between Russia and Azerbaijan and Georgia. The FPS conducted detailed checks of transport vehicles and also 'undertook special measures to uncover hidden mercenaries and fighters, their accomplices, weapons, fighting gear, devices for terror and diversion.... To defend the borders they actively used intelligence, raiding and ambush'.[19]

By 22 September all administrative borders around Chechnya were reported to be 'totally closed', as was the airspace over Chechnya.[20] The following day, the press noted that in all the regions bordering Chechnya thousands of federal troops were already stationed, constituting a true 'sanitary zone' around the republic. According to figures collected by Emma Gilligan, in all 90,000 troops were deployed to the border in addition to 30,000 MVD troops. Chechnya was sealed off.[21] There was no way of getting out, in any direction, except – for the time being – through Ingushetiya.

I have noted that the Russian leadership's re-definition of Chechnya's status from 'undecided' to an undisputable 'part of the Russian Federation' took place under cover of the substantial securitization of Chechnya as a terrorist threat, without official arguments as to why Chechnya was a part of Russia. The Russian military extended the cordon sanitaire into Chechnya and took control of the hills north of Terek in the first days of October, without any accompanying comments apart from General Manilov's statement, 'we are just deploying groups of troops to establish a security zone'.[22] It was never officially announced that a new war, with a full ground offensive, had been launched to re-take Chechnya. But the point here is that the sealing off of Chechnya as something too dangerous to be in contact with, and the multitude of security and military forces that were set to enter Chechen territory in October, had all been well grounded in official representations of the threat facing Russia. Given the construction of Chechnya as an overwhelming and dangerous terrorist threat and the resonance that this representation found among Russian audiences, the total physical isolation of the republic appeared both logical and legitimate.

Moreover, this construction stipulated violence as the *only* relevant mode of interaction with Chechnya: non-violent interaction was made irrelevant. Economic cooperation with Chechnya also shifted. On 16 September, Putin gave orders to draft plans for an oil pipeline that would bypass Chechnya.[23] On 30 September the Central Election Committee announced the impossibility of conducting December 1999 elections of candidates for the State Duma in Chechnya – justified with reference to the fact that there was no legal authority to cooperate with on Chechen territory and that the circumstances ('absence of social order') in Chechnya were such that it was impossible to guarantee the voting rights of the citizens.[24] While the Ministry of Federal Affairs and Nationalities, tasked with facilitating contact between the Federal Centre and all the different nationalities and preventing potential ethnic or religious conflicts from erupting, stopped playing any role in relations with Chechnya,[25] the different agencies empowered to administer violence, the so-called power ministries, took centre stage.

As pointed out in Chapter 4, as early as spring 1999 the security services had acquired a crucial role in Russia's dealings with Chechnya. During autumn 1999, discussions and decisions on the situation in Northern Caucasus/Chechnya were undertaken primarily by Prime Minister Putin with the heads of the power ministries.[26] Moreover, these agencies became the key 'interlocutors' in Russian–Chechen relations, dominating not only within their own sphere of competency but also those of others. The controversial decision on closing the borders between Chechnya and North Ossetia, Kabardino-Balkariya and Stavropol exactly when people were beginning to flee the intensive bombing of Chechnya in September was a direct instruction from Major General Shamanov of the Interior Forces to the Interior Ministers of these republics.[27] When refugees poured into Ingushetiya in early October and Vice-Premier Valentina Matviyenko, who was in charge of refugee issues, travelled to the region, the social and humanitarian needs of the refugees were discussed in close connection with military issues, and with the direct participation of the commanding group of the federal forces.[28]

The logical enactment of representations of Chechnya as an existential terrorist threat could thus be observed fairly immediately, in the way Chechnya was physically sealed off from the rest of Russia, as well as in the 'handing over' of all Chechen issues to the agencies that administer violence.

The physical isolation of the people

A related and difficult problematique is the way in which the securitization of Chechnya as a terrorist threat also legitimized practices that sought to seal off and 'sanitize' Russia of Chechens as such. In previous chapters, we have seen that, even if the Russian political leadership was careful not to securitize Chechens as a group, the logical sum of the discourse sometimes did just that. Moreover, an equation of Chechens with the terrorist danger quickly appeared in journalistic accounts as well as those of the experts. The question here is whether and how this implicit representation of Chechens as radically different and dangerous served to legitimize policies and practices 'beyond rules that otherwise have to be obeyed' and which targeted them as a group.

Beginning with the short war in Dagestan, practices that sought to seal off Chechens as a group quickly emerged. Akkintsy Chechens living in the Dagestani Novolak region, who had been part of the Dagestani social fabric for centuries, were not entrusted with weapons to help fight back the invaders. This clearly was a change in the rules which had guided societal life in Dagestan.[29] Moreover, Akkintsy Chechen refugees fleeing from these regions to Khasavyurt did not receive any help from the administration. According to these refugees, the police detained and beat up innocent young Akkinsty Chechen boys without reason.[30] Finally, when Chechnya was being bombed in late September, MVD forces and police guarding the border between Dagestan and Chechnya did not allow Chechen refugees into Dagestan, only native Dagestani ethnicities such as Avars, Dargins and Nogais, and Slavs.[31]

Such filtering out of Chechens as an ethnic group was also practised in other neighbouring republics. At the end of October, for example, three Chechen football players on the Ingush football team 'Angusht' were detained on the border to Stavropol kray and could not take part in the match against the Rostov team Avtodorom.[32] According to media reports, detaining persons of Chechen origin, or holding Chechen passports, on the border quickly became widespread.[33] From August 1999, the authorities of North Ossetia as well as Kabardino-Balkariya prohibited entry for all Chechens, irrespective of the region of their permanent registration according to place of residence.[34] These must surely be seen as measures beyond the rules that otherwise have to be obeyed as they explicitly broke with Constitutional provisions to protect against discrimination on the basis of nationality.[35] The practice of using registration requirements to deport Chechens, discussed in detail in the Moscow case below, was also widely adopted in Krasnodar, Stavropol, Kabardino-Balkariya and

North Ossetia and even in Krasnoyarsk and Volgograd.[36] In Volgodonsk, the city in Rostov oblast struck by a terrorist attack on 16 September, there were repeated calls for deporting all Caucasians from the city. There were also several cases of Chechens being beaten up by people living in Volgodonsk after the terrorist attack.[37]

Such emergency measures, which served to seal off Chechens from the neighbouring regions, became fairly widespread, and were clearly beyond rules that otherwise have to be obeyed, in social as well as legal terms. They were, however, fully in line with the securitization of Chechens as a dangerous group of people.

Practices that equated 'Chechen' with the terrorist threat and resulted in sealing off Chechens from Russia were evident not only in the border regions, but also across Russia. Moscow Mayor Yury Luzhkov had promised the citizens of Moscow that 'harsh *(zhëstkiye)* and radical measures' would be taken after the latest bomb explosion in Moscow: 'all those we cannot be sure of *[te v kom my ne mozhem byt uvereny]* will be expelled from Moscow'.[38]

On 13 September, Luzhkov issued Order no. 1007 ('On immediate measures to establish order in the registration of citizens temporarily residing in Moscow'), which required the deportation of non-registered people from the capital. This was accompanied by Resolution No. 875, published by the government of Moscow on 21 September, 'On the approval of the temporary order of movement of persons who are violating the rules of registration, out of Moscow to the place of their residence', which sanctioned the deportation of those without permanent residence in Moscow. The Order and the Resolution were in contravention not only of international conventions signed by Russia but also of key provisions in the Russian Constitution as well as other governing legislation on protecting freedom of movement.[39] The main problem with the Order was that registration, according to the Constitution, was intended as a system of simple notification when moving or changing a place of residence. Now a system was introduced that required, in practice, Russian citizens to have formal permission to stay in Moscow. The deportation of Russian citizens in the given situations was also in breach of Russian law.[40] The Order and the Resolution were not only illegal in content: they were also issued in an illegal manner. They were not published, and thus there were no legal grounds for their implementation.[41] On 28 September, when the refugee flows out of Chechnya had reached unprecedented levels, yet another Order, No. 1057 ('Temporary measures for systematizing work with refugees and forced migrants arriving in Moscow, as well as with persons who apply for the corresponding status'), was issued but not published by the Moscow authorities. It too was in contravention of Russian laws[42] and sanctioned practices that served to seal off Moscow from Chechens.[43]

These new Orders definitely moved 'beyond the rules' in legal terms. This is not to say that certain groups (of Russian citizens) in Moscow had not had their rights violated in connection with registration requirements previously[44] – but the securitization of the Chechen terrorist threat was so far-reaching that it enabled the adoption of new legal codes in clear breach of Russia's legal foundations.

The extraordinary regime (under Order no 1007) introduced on 13 September by the Moscow City Government encountered only very limited opposition in the Duma. A small group of independent deputies including Sergey Kovalyev, Sergey Yushenkov and Viktor Pokhmelkin moved that the Duma should consider a resolution on 'the necessity of compliance with the constitution and the laws of the Russian Federation during the implementation of counterterrorist activities', but this was not supported by the majority (62 for, 136 against).[45] Not only did the new directives contravene the legal foundations of Russia – they also seemed to break with core societal rules. Russia and Moscow in particular have always been considered multi-ethnic and multi-confessional. Now practices that systematically, and on a large scale, infringed the rights of certain groups had become acceptable to society.[46]

Equally pertinent is how the practices stipulated in these Orders enshrined the linguistic representation of the terrorist threat as extremely dangerous. Those Russian citizens who were subjected to the new procedures were classified as potential terrorists; thus, according to Order No. 1007 for example, they had to be 're-located out of Moscow to the place of their permanent residence', if registration requirements were not met. Moreover, they were so dangerous that 'before the re-location to the places of permanent residence the persons, subject to moving out, should be kept at the militia stations.... The same Departments of Interior have to send militia officers to escort the deportees'.[47] These directives stipulated practices that were fully congruent with the threat level implied in the dominant representations of the Chechen terrorist threat that autumn: and they were not empty words on paper.

In line with the new Order 1007, 'Operation Foreigner' was launched in Moscow from 14 September 1999 to cleanse the city of unregistered persons by forcing non-residents of the city with short-term permits to re-register.[48] In theory, the procedure should have been applied to anyone without the required documents for registration, but in practice the people 'we cannot be sure of' (in Luzhkov's words), were from the Caucasus, Chechens above all, even when they had the complete set of documents required for registration.[49] Zaynab Zadulayeva, for example, a Chechen mother of four, who had lived in Moscow for two years, was refused re-registration, on the grounds that she was staying with a friend and not a relative, as noted in her registration. The police officer took her passport, tore up her registration and shouted 'Be off, or I will call the OMON (police)'.[50] By 29 September, 19,000 non-residents had been denied registration and 10,000 non-Muscovites had been deported from the city.[51] When the intensive bombing of Chechnya started in October, Order No. 1057 was put to use to prevent fleeing Chechens from entering Moscow and staying there.[52]

Despite these restrictions, many Chechens still settled or continued to live in Moscow, often without the required registration. But they were also 'sealed off' from Russian society. Registration became a precondition for the exercise of basic rights and freedoms such as employment, marriage registration, participation in elections, medical care, pensions and allowances, secondary and higher education.[53] In sum, Chechens were either removed from Moscow, or were sealed off from all normal activities and encounters with other members of this society. In turn, these material expressions of the new dominant representation of the Chechens served to reinforce the construction of Chechens in Moscow as 'different' and 'dangerous'.

Over time the practice of using registration requirements to cleanse Moscow of Chechens became routine.[54] It intensified following incidents such as the 2002 terrorist attack at the Nord-Ost theatre in Moscow.[55] Such peaks were justified by re-articulations of Chechens as an existential threat to Russia. According to a journalistic account from *Moskovsky Komsomolets*, for example:

> The way to conquer our fear of Chechens is simply not to let them into Russia.... Our true target should be to restrict the rights and freedoms of Chechens as representatives of a people with whom we have been at war for a long time. Whichever way you look at it they represent a potential threat to the safety of our children, and we should not close our eyes to this fact.[56]

Securitizing language not only enables the undertaking of emergency practices in the first place: it also legitimizes their continued use. New waves of such talk contribute to reify and uphold the identification of Chechnya and Chechens as different and dangerous, just as the material enactment of this talk over time does.

Other practices which became widespread in and beyond Moscow, legitimized by the representation of Chechens as different and dangerous, included illegal checks and detentions, often resulting in fabricated criminal cases being brought, usually on charges of carrying illegal weapons and/or drugs. These practices, which were in evident contravention of the Russian Constitution as well as a whole set of other Russian laws, were not given any special legal framework. They were not entirely new and had been used against people of other nationalities as well.[57] However, the securitization of the terrorist threat in autumn 1999 took this practice to new heights and legitimized it. And Chechens seemed to be targeted in particular.

It is difficult to judge the extent of this practice, but it appears to have become fairly widespread in the immediate aftermath of the 1999 bombings in Russia. Special operations under the label of 'Whirlwind Anti-Terror' were planned in the republics bordering Chechnya and in key Russian cities in mid-September.[58] Only two days after this anti-terror operation was announced, newspapers reported that in Moscow 2,200 wanted persons had been detained and more than 9,000 persons suspected of taking part in criminal acts had been detained.[59] In St Petersburg, 16,000 police officers took part in the operation; as early as 22 September came reports that 1,463 crimes had been revealed.[60]

Judging from the reports of human rights organizations, illegal detentions were often the result of these campaigns, which were most frequently undertaken against Chechens, but also against Dagestanis and Azeris, in autumn 1999.[6] Memorial reported that 'mass fabrication of criminal accusations' against Chechens accompanied these detentions, and concluded that 'as a rule most o the arrested are found guilty in the courts'.[62] The example of Ruslan Musitov is but one.

Ruslan was a resident of Grozny and the Deputy Chairman of the Chechen Department of the International Human Rights Society. He came to Moscow on 22 September to attend the Congress of the Otechestvo political movement and was put forward as a State Duma candidate for the 37th electoral district. On 27 September, employees of the District Department for Fighting Organized Crime requested Musitov to step out of the flat where he was staying and into the street. There they allegedly found two matchboxes with drugs and three bullets. Musitov was detained for three days. Then the term of his detention was prolonged for ten days; thereafter he was transferred to investigation prison No. 2.[63]

As with registration requirements, illegal detentions and the fabrication of criminal charges against Chechens were not limited to the big cities. According to Memorial, illegal detentions and the fabrication of criminal charges also took place in Krasnodar, Stavropol, Volgograd, Nizhny Novgorod, Tomsk and Rostov.[64] These practices became routine over time, and rose to new levels following incidents such as the 2002 terrorist attack at the Nord-Ost theatre in Moscow. According to Memorial, some 400 Chechens were detained throughout the Russian Federation in the days following that attack. Scores of Chechen male residents in Moscow were picked up for routine identity checks and charged with weapons or drugs possession.[65]

Once again, these practices were nothing new in Russia, but the heavy discourse on Chechnya as an existential terrorist threat made their mass-scale application appear both logical and legitimate. The illegal detention of a Chechen became all the more acceptable when this Chechen was constructed as different and dangerous. As can be seen from the dialogues in the two stories presented below, even the language accompanying the execution of these practices on the micro-level was informed by the discourse on Chechnya/Chechens as an existential terrorist threat to Russia.[66]

1   Irina, the wife of a detainee Badrudi Eskiyev spent all of 15 September searching for her detained husband. She found him at the Pechatniki Department of the Interior. She was sent to room 503 where a man in civilian clothes was sitting.

> IRINA'S STORY: 'I tell him: 'I need Eskiyev. A man is lost. Where is he?' He answers, 'Probably in jail.' 'How? Why?' 'He is a Chechen, he probably smokes grass, takes drugs. All Chechens are like that.' 'How can you say such things?' 'And how is it possible to blow up people's homes?' 'If somebody does that, it does not mean that the whole of the people have to be blamed.' Then he says, 'A good Chechen is a dead Chechen. All Chechens have to be killed.' I started to cry, and said. 'You are wrong.' And he said to me, 'Go away, we shall be discussing that for a while. Come back in three days.' In the morning of 16

September, at the Tekstilshchiki Department of the Interior, the investigator Avdeyeva declared to Ira that her husband had been detained in Tekstilshchiki Street and drugs were found in his pockets, in connection with which a case had been taken out against him under Article 122. In reality, Badrudi had been detained early in the morning at home, taken out of bed by the militia, put on thoroughly checked clothes in which nothing was found, and taken away before his wife's eyes.

2   An employee of the Chechen fiscal police was on a business trip to Moscow. 'One day all my documents became invalid. It happened on the 14th. My driver and I were taken to the Regional Department for Combatting Organized Crime. I was trying to find out what had happened, how I had violated the law, and they were saying: 'All of you Chechens are our enemies, you are attacking our homes.' I said again, 'what specifically do you have against me and what does my ethnicity have to do with it?' He then said: 'And you are working, receiving money and then sending it to guerrillas!' and then returned the documents but with the warning: 'if you do not leave, if we meet you again in Moscow, we will find drugs in your pockets, as well as explosives and armaments'. And my driver was beaten and called a Chechen. Now it is as if I am under house arrest. I do not leave my home without utter necessity. We now have no rights, unprotected and needed by nobody. Such hunting for us never happened before. Whoever you call – every second one has a story to tell'.

The representation of the Chechen as different and dangerous accompanied the execution of this practice and served to legitimize the illegal actions these Russian servicemen were undertaking there and then. It would be wrong to think that the securitization of Chechnya and Chechens was produced from the top of the political system, with input from the 'side' but not from 'below', that is, from the ordinary talk about Chechnya in Russia. When the Moscow Helsinki Group opened a 'hotline' to give legal advice to people in Moscow who were facing re-registration demands by police in connection with 'Operation Foreigner' following the bomb explosions in September 1999, calls from Muscovites dominated the line during the first two days. Apparently, what they had to say went along the lines of 'why did you open this line, we need to clean Moscow of these people. They will not let us live', or 'Why do we defend people from the Caucasus, they always behave like a mob. It is necessary to throw them out of Moscow'.[67] The point here is not to claim that these callers are representative of all Muscovites in general, but to show that the securitizing narrative launched from the political leadership resonated well with and probably served to accentuate a discourse on Chechnya/the Caucasus already existing among the Russian police and the population.

The statements above also clearly draw on the dominant position articulated by the Russian political elite, experts and journalists that autumn. Whatever the intersubjective status of these linguistic transactions, the reiteration of the dominant position on Chechnya in everyday speech helped to solidify this construction with yet another layer, anchoring it at this lower level of Russian society.

Conclusion

Starting with the endorsement by the Russian Federal Assembly of the entire 'plan' presented by Prime Minister Putin on how to deal with Chechnya in September 1999, most explicit suggestions by the Russian leadership on how to fight the terrorist threat were endorsed by what could have constituted a political opposition in the chambers of the Russian Federal Assembly. This pattern of endorsement of measures 'beyond rules that otherwise have to be obeyed' in connection with the counter-terrorist operation did not change over the years. On 21 November 2002, for example, a controversial and unconstitutional bill forbidding the government to return to their families the bodies of suspected terrorists killed during counter-terrorist operations and/or to reveal where those bodies were interred was supported by 296 deputies in the Duma, with only 34 voting against and four abstaining.[68]

In an account such as this, the emergency measures that were *not* explicitly outlined in official policies, plans or laws, but were still enabled by the securitizing narrative are even more important than such formal endorsement. In my view, the lack of a clear legal foundation and explicit codes of conduct during the counter-terrorist operation made the various components in the securitizing narrative particularly important: the representations of the threat and the advice on the 'way out' implicit in statements by the Russian political leadership and Russian generals stood alone as 'instructions' on how the war could be fought. The practices that the securitizing narrative legitimized and how this worked have made up the backbone of this chapter, as they will for the next two.

This chapter has shed light on two key practices enabled by securitizing talk that were undertaken against Chechens in Russia outside of Chechnya.[69] Taken together these practices served to seal off Moscow and other cities from Chechens, which seemed very necessary given the new dominant representations of this group. Just as Chechnya itself was being sealed off with high ditches, closed borders and thousands of troops around its borders, these practices resulted in the physical and social isolation of Chechens from mainstream Russian society. In line with the re-articulation of the Chechens as an existential terrorist threat to Russia, they could be detained, deported and prevented from crossing the border. They were most logically placed in police stations or in jails, not in schools, workplaces, football fields or other public arena. Just as any interaction with Chechnya was undertaken by the agencies that administer violence, and not by those responsible for inter-ethnic or humanitarian issues, the Chechens were now most logically dealt with by security personnel.

I have argued that the terrorist scare built up through linguistic representations during autumn 1999 made possible the adoption of emergency measures 'beyond rules that otherwise have to be obeyed' both legally and socially, in the sense that such measures now could be undertaken legitimately and on a massive scale, although they were not totally new. The enactment of such practices 'on the ground' was accompanied by linguistic representations that were linked to and referred to the dominant discourse of Chechnya as an existential terrorist threat. Not only did these new re-iterations reinforce the dominant discourse: they also served to carry it further into the future and spread it downwards into Russian society. We have also seen that repetitions of these practices over time have been accompanied by new articulations of Chechen identity, so as to legitimize their continued enactment. New peaks in detaining or re-registration practices do not happen out of the blue or in silence: they are carried on new waves of securitizing talk.

Finally, the material enactment of linguistic representations *over time, in visible practices* served to reinforce the construction of Chechnya and Chechens as different and dangerous. The physical sealing off of Chechnya from Russia and Chechens from Moscow, and the daily and continuous handling of this territorial unit and this group mostly by security personnel, served to reify their identity as different and dangerous. 'Is someone being deported or detained?.. Aha, must be a Chechen'.

Notes

1 B. Buzan, O. Wæver and J. de Wilde. 1998. *Security: A new framework for analysis.* Boulder, CO: Lynne Rienner, 25.

2 Several key decisions taken for dealing with Chechnya as a security threat did not require formal endorsement by the Federal Assembly because power is highly concentrated in the president in the Russian political system. Presidential decrees have been widely used and, apart from those concerning emergency and military regimes, do not need formal endorsement, despite the wide consequences they may entail.

3 *RFE/RL Newsline*, 15 September 1999 and 'Putin predlagayet novyy plan chechenskogo uregulirovaniya', *NeGa*, 15 September 1999.

4 In January 1997, there was strong support (67 per cent) for the Khasavyurt Accord which stipulated the withdrawal of Russian forces from Chechnya (B.K. Levashov 2001. *Rossiyskoye obshchestvo i radikal'nye reformy*. Moscow: Akademia, Russian Academy of Science, Institute of Social-political Research, 851).

5 The first step in the final point of the plan, 'the creation of a Chechen government in exile', was a decision to appoint a presidential plenipotentiary to Chechnya; it took the form of a presidential decree issued on 15 October (Presidential Decree no 1380 posted in *RoGa*, 19 October 1999). Without stating so explicitl, this decree discounted Maskhadov's status as the legitimately elected president of Chechnya.

6 'Chechenskiy syuzhet ne dolzhen povtorit'sya v Dagestane', *Parlamentskaya Gazeta* 18 September 1999 and 'Vystupleniye Putina ponravilos' senatoram', *NeGa*, 1 September 1999.

7 'Putina', *Profil*, 27 September 1999.

8 'Skazochnik s kholodnymi glazami', *Moskovskiy Komsomolets*, 16 September 1999.

9 Ibid.

10 'Putin predlagayet novyy plan chechenskogo uregulirovaniya', *NeGa*, 1 September 1999.

11 The 1998 'Law on Combating Terrorism' defines such an operation as 'special activities aimed at the prevention of terrorist acts, ensuring the security of individual

neutralizing terrorists and minimizing the consequences of terrorist acts'. In other words, it seems aimed at suppressing a specific act of terrorism in a limited zone and over a limited timespan. The law significantly expands the categories of officials with a law-enforcement mandate and does not specify the circumstances under which fundamental human rights may be curtailed, or the degree to which they may be restricted (Federal Law on Combating Terrorism, enacted 25 July 1998, articles 3, 10, 6 and 7).

12 'Vokrug Groznogo szhimayetsya kol'tso', *NeGa*, 28 October 1999.

13 *RFE/RL Newsline*, 30 September 1999.

14 Interview published in 'Zakon na storone Federalov', *NeGa*, 13 October 1999.

15 'Moskva prinimayet bespretsedentnyye mery po bor'be s terrorizmom', *NeGa*, 18 September 1999.

16 'Na Kavkaz otpravilsya samarskiy OMON', *NeGa*, 1 October 1999.

17 General Valery Manilov cited in 'Taynyye i yavnyye manevry Moskvy', *NeGa*, 5 October 1999.

18 'V Stavropole usileny mery bezopasnosti', *NeGa*, 18 September 1999.

19 According to official sources, between 1 September and 28 October, 190,000 people were questioned – using 'special methods' with 3,700 of them. This resulted in the uncovering of 34 persons suspected of belonging to terrorist organizations, the capture of 570 illegal immigrants, the expulsion of more than 500 people and the handing over of approx. 400 wanted persons to the FSB and MVD. ('V Dagestane po-prezhnemy nespokoyno', *NeGa*, 29 October 1999).

20 'K novoy voyne v Chechne pochti vsë gotovo', *NeGa*, 23 September 1999.

21 E. Gilligan. 2010. *Terror in Chechnya: Russia and the tragedy of civilians in war.* Princeton, NJ: Princeton University Press, 34.

22 'Taynyye i yavnyye manevry Moskvy', *NeGa*, 5 October 1999.

23 *RFE/RL Newsline*, 17 September 1999.

24 'Zayavleniye Tsentral'noy izberatel'noy komissii Rossiyskoy Federatsii o nevoz-mozhnosti podgotovki i provedeniya vyborov deputatov Gosudarstvennoy Dumy Federal'nogo Sobraniya Rossiyskoy Federatsii tret'yego sozyva na territorii Chechn-skoy Respubliki 19 Dekabrya 1999 goda', announced in *RoGa*, 13 October 1999.

25 According to the North Ossetian President, the Ministry of Federal Affairs and Nationalities had been turned into an 'agency escorting the force ministries in their travels around North Caucasus'. ('Minnats popal pod ogon' kritiki', *NeGa*, 24 September 1999). The Ministry of Federal Affairs and Nationalities was later abolished by the Presidential Decree of 16 October 2001.

26 'Putin provel soveshchaniye silovikov', *NeGa*, 24 September 1999.

27 Memorial and Civic Assistance. 1999. *The report on the observer mission to the zone of the armed conflict, based on the inspection results in Ingushetia and Chechnya*, 3. Available at www.memo.ru/eng/memhrc/texts/ch2.shtml and accessed 17 March 2016.

28 'Matviyenko priyekhala k bezhentsam', *NeGa*, 7 October 1999. The Ministries of Defence, the Interior, Justice, as well as the Federal Security Service (FSB) were all joined under the Unified Group of the Russian Federation Armed Forces (OGV) of the Northern Caucasus.

29 R.B. Ware and E. Kisriev. 2010. *Dagestan: Russian hegemomy and Islamic resist-ance in the North Caucasus.* New York: M.E. Sharp.

30 'Voyna posle Voyny', *NeGa*, 17 September 1999.

31 'Potok bezhentsev narastayet', *NeGa*, 29 September 1999.

32 'Grozny budut brat' po chastyam', *NeGa*, 22 October 1999.

33 'Podozritel'nykh lovyat na yuzhnoy granitse', *NeGa*, 22 October 1999.

34 Memorial. 2000. Compliance of the Russian Federation with the Convention on the Elimination of all Forms of Racial Discrimination, 22. Available at www.memo.ru/hr/discrim/ethnic/disce00.htm and accessed 17 March 2016.

35 According to the Constitution Article 19 (2):

> The state shall guarantee the equality of rights and liberties regardless of sex, race nationality, language, origin, property or employment status, residence, attitude to religion, convictions, membership of public associations or any other circum stance. Any restrictions of the rights of citizens on social, racial, national, lin guistic or religious grounds shall be forbidden.

This includes all basic human rights, freedom of movement and of residence, protec tion by the law, assumption of innocence, etc. (Article 4(2)).

36 *RFE/RL Newsline*, 24 September 1999 and Memorial (2000).

37 'Ekho vzryva na tikhom Donu', *NeGa*, 26 October 1999.

38 'Protiv ChP vystopayut vse', *NeGa*, 14 September 1999.

39 According to the decision of the Moscow City Court of 25 September 2000, the docu ments (Order 1007-PM and Resolution 875) were issued in violation of the Constitu tion (Article 27 (1) – freedom of movement, Article 55 (3) – prohibition of unlawful limitations of human rights) and federal legislation (Law of Russian Federation 'On the Right of Russian Citizens to Freedom of Movement, the Choice of a Place to Stay and Reside within the Russian Federation' (1993)), Articles 3 and 8 of this law.

40 According to the decision of the Moscow City Court of 25 September 2000. Under the Code of Administrative Offences RSFSR (1984), there was simply no such sanc tion (for any administrative offences) for deporting Russian citizens to their region o permanent residence. Hence, deportation according to Resolution 875 was an illegal sanction.

41 According to both Article 15 (3) of the Constitution and the Charter of the City o Moscow (Article 10).

42 It contradicted the federal law 'On Refugees' (1993), Article 5 and the federal law 'On Forcibly Displaced Persons' (1993), Article 6. In fact, the Directive 1057 was repealed by the Supreme Court of Russia in 2001, for contravening Russian legislation.

43 See Olga Cherepova in Memorial 1999a, *Moscow after the Explosions. Ethnic cleans ings. September–October 1999*, 2. Available at www.memo.ru/eng/hr/ethn-e2.htm and accessed 31 July 2013.

44 Memorial recorded such practices in Moscow in the years before 1999 (Memorial 1999b. *Ethnic Discrimination and Discrimination on the Basis of Place of Residence in the Moscow Region*. Available at: www.memo.ru/eng/hr/ethn-el.html).

45 'Vikhr'-antiterror dayet polozhitel'nyye rezul'taty', *NeGa*, 28 September 1999.

46 'Do you agree with the regime of registration becoming more strict?' Yes: 93.7 pe cent, no: 7.9 per cent, it is none of my business: 1.9 per cent (Poll referred to in Memorial 1999a, 8).

47 Quoted in Memorial 1999a, 2.

48 As early as August and accompanied by the information from the Russian Security services that 'diversionists from Chechnya are preparing to carry out terrorist acts in all Russian major cities', the MVD informed that it was preparing to 'cleanse places *kompaktnogo prozhivaniya* (densely populated) by Caucasians' ('Seyat' uzhas smert' v rossiyskikh gorodakh', *Kommersant*, 11 August 1999).

49 *RFE/RL Newsline*, 23 September 1999. According to Olga Cherepova newcomers were

> registered selectively, with almost all Russians receiving registration, while many Azeris, Armenians, Georgians and others arriving from the Transcaucasian Republics and Northern Caucasus are refused; all Chechens are refused, even i there is a complete set of documents required for registration.
>
> (Memorial 1999a, 3)

50 'V Rossii vsekh propishut', *NeGa*, 22 September 1999.

51 *RFE/RL Newsline*, 30 September 1999 and Memorial 2000.
52 Memorial 1999a, 3–4.
53 Memorial 2000, 3–4.
54 Svetlana Chuvilova, who ran a telephone hotline for Civic Assistance, responded in October 2000 to a question about Chechens being denied registrations that she received over 100 registration-related complaints every week ('In Moscow people complain of racial profiling', *Christian Science Monitor*, 27 October 2000). In 2003, Amnesty International could still report that most Russian nationals subjected to registration problems and expulsion from Moscow were Chechens (Amnesty International. 2003. *Rough justice: The law and human rights in the Russian Federation*. London: Amnesty International Publications, 40–45. Available at www.amnesty.org/en/library/info/EUR46/054/2003/en.
55 Amnesty International, 2003.
56 'Terror. Gadky privkus svobody', *Moskovskiy Komsomolets*, 30 October 2002.
57 Memorial 1999b. *Ethnic Discrimination and Discrimination on the Basis of Place ofRresidence in the Moscow Region*. Available at www.memo.ru/eng/hr/ethn-el.html and accessed 31 17 March 2016.
58 'Ekstremisty ob''yavili Rossii otkrytyy terror', *NeGa*, 16 September 1999.
59 'Moskva prinimayet bespretsedentnyye mery po bor'be s terrorizmom', *NeGa*, 19 September 1999.
60 'V Piterskom obshchezhitii progremel vzryv', *NeGa*, 22 September 1999.
61 Some police officers admitted to have been given 'verbal orders based on a directive' from the Chief of the Moscow Head Department of Internal Affairs to 'detain and not re-register Caucasians, and primarily Chechens (Memorial 2000, 11).
62 Memorial 2000: 9–10.
63 Memorial 1999a: 4–5.
64 Memorial 2000: 6–7.
65 Memorial referred to in Amnesty International 2003, 46.
66 Both stories are taken from Memorial's collection of evidence of abuse following the bombings in September and October 1999 (Memorial 1999a, 5 and 7).
67 Cited in 'Goryachaya liniya', *NeGa*, 13 October 1999.
68 'Duma agrees to "wage war with corpses"', *RFE/RL Newsline*, 21 November 2002.
69 The practice of refusing to issue passports (for travel abroad) to Chechens is another example (Memorial 2000, 10).

11 Bombing Chechnya

This chapter turns to the war zone proper. Outlining the continuous bombing of Chechen territory from early September 1999 until early 2001, it argues that these bombing practices went beyond both legal and social rules. Language is invoked to understand how these practices nevertheless became acceptable. The ground had been well prepared in linguistic representations for a new war against Chechnya. However, the focus is on how linguistic and material practice *worked together* to strengthen the discourse on Chechnya as an existential terrorist threat, adding ever-new layers and making the war acceptable even as it unfolded in all its cruelty. 'Emergency measures' are therefore again studied by exploring the link between two aspects: the linguistic representations in the securitizing narrative and implementation of these in bombing practices which then again serve to constitute and maintain the linguistic identity construction in the securitizing narrative.

This chapter also investigates the discursive handling of potentially 'shocking events' on the battlefield (such as gross human rights violations or the killing of civilians). Despite an emerging media blockade on events in Chechnya, news of particularly violent incidents during the Second Chechen War did enter Russian public space. As securitization is never a stable social arrangement, the exposure of such potentially shocking events as the war unfolded could have prompted return of alternative positions on 'Chechnya' and 'Russia' in Russian discourse, even a re-emergence of the 'discourse of reconciliation'. Official statements, as well as those of Federation Assembly members and experts and journalist accounts of such events in the war, will therefore be investigated in this and the next chapter. A key argument throughout these two chapters will be that such shocking events were continuously 'carried' and 'covered' by references to the initial securitizing narrative.

A campaign beyond the rules

In the war zone itself, the physical sealing off of Chechnya from Russia was followed by a massive bombing campaign. From early September, Chechnya was bombed, without any prior communication with Chechen authorities. Beginning late on 17 September 1999, Russian aircraft flew some 100 raids on

Chechen targets in the course of 24 hours, and continued bombing the following day.[2] On 23 September, suburbs of Grozny and the airport were under bombardment.[3] By the end of September the entire territory of Chechnya was being bombed: industrial areas, oil wells and installations, roads and bridges as well as residential areas.[4] From the beginning of August to the end of September, between 1,250 and 1,300 bombing raids were carried out over Chechen and Dagestani territories, according to official Russian sources.[5]

Even if the main air campaign was over by October and the focus of military activities shifted to the installation of a cordon sanitaire in Northern Chechnya, the bombing of Chechen territory continued. To avoid the heavy casualties suffered by the federal forces during the first Chechen conflict, this time a 'minimum risk approach' was employed – which meant sending in infantry only after heavy artillery and air bombardment had been carried out.[6] When Grozny was encircled in late October, the Russian press reported that 34 bombing raids were carried out across Chechnya every day, hitting Gudermes, Grozny and the surrounding hills in particular.[7] On 28 October, these bombardments were increased. According to official sources, over 100 aircraft bombing raids over Chechnya were carried out in 24 hours, accompanied by heavy artillery bombardment.[8] Grozny, subjected to a constant bombing campaign throughout November, was finally totally 'blocked' in the beginning of December.

Following statements by Russian intelligence that there were still 2,000 'terrorists' left in the city, flyers were distributed on 6 December, demanding that everyone leave the city before 11 December: otherwise they would be 'annihilated'.[9] The problem was that as many as 20,000 to 40,000 people were still in Grozny, many of whom could not leave the city or simply did not know about the ultimatum.[10] When the bombing resumed, not on 11 December, but as early as 7 December, thousands were trapped inside the city. In the weeks that followed, Grozny was subjected to a mass and indiscriminate bombardment that, according to the Russian military analyst Pavel Felgenhauer, hit the civilian population much harder than the rebels.[11] Similarly, thousands of civilians were trapped in the southern villages which also were subjected to intensive shelling during these weeks. By all reasonable estimates, the bombing campaign of Grozny was much heavier than it had experienced during the First Chechen War. This time weapons such as the tactical missiles 'Tochka-M', 'Tyulpan' mortars and aviation bombs weighing 2.5 tonnes or more were used.[12]

The aim here is not to offer exact figures, but to give an impression of the massive scale of the bombing. According to Marcel de Haas, between October 1999 and February 2000, airpower was used in more than 4,000 combat sorties, of which the majority were strike sorties.[13] Even after Grozny and most of Chechnya had finally been recaptured in February 2000, the battle for the southern mountains continued at least till the beginning of 2001, conducted with the help of Russian Air Forces bombing Chechen positions there.

In sum, then, a massive bombing campaign was carried out over a lengthy period, targeting the entire Chechen territory. According to Gilligan, the huge number of refugees entering Ingushetiya (250,000) testifies to the scale of the bombing campaign.[14] Because of the breadth of the strikes across the region, south, west and east of the capital, civilians could not seek refuge in the countryside as they had done during the First Chechen War, but flooded into Ingushetiya. Nor were the bombing raids mainly pinpointed strikes targeting 'places where terrorists were concentrated' to 'minimize the casualties amongst the civilian population', as claimed by Commander of the Western group of Federal Forces in Northern Caucasus Vladimir Shamanov.[15] There were many civilian casualties and several documented cases of indiscriminate bombing.[16] Even civilian convoys were bombed.[17] Moreover, the repeated closing of border checkpoints into Ingushetiya by Russian military personnel meant that refugees could not get out while the bombing was underway.[18] There is also evidence of extensive use of illegal ammunition. TOS-1 Buratino 30-barrel multi-rocket launchers were used: these are air-delivered incendiary weapons intended to set fire to objects or cause burn injuries to those on the ground.[19] Memorial reported on the use of cluster bombs in a 'carpet bombing of the village of Elistanzhi' as early as the beginning of October.[20] Cluster bombs were also used in the 21 October bombing of the Grozny Central Market discussed below.[21]

That such a bombing campaign was 'beyond rules that otherwise have to be obeyed' in legal terms has been determined in several decisions of the European Court of Human Rights. In its judgments related to the events in Chechnya during autumn/winter 1999/2000, the Court concluded that the operations were planned and executed without due care for the lives of the civilian population, and in violation of the Article 2 (right to life), among other articles, of the European Convention on Human Rights and Fundamental Freedoms.[22] In its judgment related to the attack on the village of Kogi in September 1999, the Court stated that it was 'struck by the Russian authorities' choice of means in the present case for the achievement of the purpose indicated' (para.147). The Court concluded that the village, in fact, 'came under indiscriminate bombing by federal air forces' (para. 148), and that the attack was 'manifestly disproportionate' (para. 150). The decision on Isayeva, Yusupova and Bazayeva v. Russia, concerning the bombing of a civilian convoy in October 1999 also referred to disproportionality. The Court pointed to the excessive use of force in stating that '[t]he military used an extremely powerful weapon for whatever aims they were trying to achieve' (para. 195). Furthermore:

> even assuming that the military were pursuing a legitimate aim in launching 12 S-24 non-guided air-to-ground missiles on 29 October 1999, the Court does not accept that the operation near the village of Shaami-Yurt was planned and executed with the requisite care for the lives of the civilian population.
>
> (para. 199)

Similar issues about proportionality have been addressed by the European Court of Human Rights in several other cases as well.[23]

This was a bombing campaign that also went beyond the rules in *social* terms. Western leaders immediately raised their voices, protesting that these bombings could not pass for counter-terrorist measures.[24] In the Russian social context also, these were measures 'beyond the rules'. As noted, during the First Chechen War there had been protests in Russia against such massive and indiscriminate bombing: it is not as if Russian society always and necessarily accepts this kind of massive violence. Despite the media blockade that was eventually put in place, news of the bombing campaign as well as of civilian casualties was to some extent covered in Russian media during autumn/winter 1999–2000. How could such massive violence against a civilian population and a territory held to be part of the Russian Federation be acceptable?

How language matters I

The argument I am advancing is that the ground had been prepared in linguistic representations of Chechnya/the Chechens that autumn. This massive (and at times indiscriminate) violence was logical and legitimate, given the dehumanized, overwhelming and dangerous nature ascribed to this territory and indirectly to this group of people during autumn 1999, both in official representations and in those of politicians, experts and journalists. Putin's promise that 'the bandits will be destroyed wherever they are', as well as the equation of the Chechens with 'terrorists' and ideas of the Chechens as collectively guilty for the terrorism that had been unleashed against Russia, contributed to legitimize massive bombing and even the bombing of civilian targets from the very beginning of the war.

Again, I am not claiming that this material practice of subjecting Chechnya and Chechens to massive violence was *caused* by linguistic representations of Chechnya and Chechens, nor that it was a *new* practice in Russia's relations with Chechnya. Massive bombing, indiscriminate bombing and the use of cluster bombs had been a prime feature of the First Chechen War as well; and massive violence and a lack of concern for civilian casualties had characterized Russian warfare in the Caucasus 200 years previously.[25] The point here is that a new and specific instance of the general practice of massive bombing was foregrounded in and legitimized by linguistic representations that attached a similar level of threat to the object as the level of violence employed against that object (on scaling of threat, see Chapter 2). This congruence made practices, such as massive bombing of Chechnya, appear logical as well as legitimate.

Further, as we will see, linguistic representations of the threat *before* going to war did more than legitimize massive bombing at the outset. Ever-new bombardments were continuously legitimized by new linguistic articulations. This served to anchor these practices in the securitizing narrative that was offered to legitimize them in the first place, making such practices appear both logical and legitimate even as the high human cost of war became evident.

When news of civilian casualties emerged, Russian officials often denied that the incident had taken place at all – as when a bus filled with refugees was hit on

5 October. Defence Minister Igor Sergeyev and the armed forces' first deputy chief of staff Valery Manilov said they had no information about an attack and they would have been informed had such an incident taken place. Sergeyev called the report 'disinformation'.[26] According to Prime Minister Putin, such information on the bombing of peaceful civilians in Chechnya was merely the 'ill-natured propaganda of the terrorists'.[27] Other times potentially 'shocking events', such as the indiscriminate bombing of the village of Elistanzhi on 7 October, were not commented upon by Russian officials at all, not even to deny them.[28] Instead, they were camouflaged by the continuous discourse on Chechnya as an existential terrorist threat. On 13 October, the press agency of the Ministry of Defence provided information about how 'concentrations of fighters' had been 'destroyed' in several parts of Chechnya, how the 'Chechen extremists' might build armed formations of up to 25,000 consisting mostly of young people under the age of 18, how they were planning new terrorist acts, also against nuclear facilities, and how they were provoking federal forces to strike at civilian targets.[29] If the word 'Elistanzhi' figured in such accounts, it was merely as one of many places where these 'bases of the band formations had been destroyed'.[30]

If, however, civilian casualties were admitted as a reality and commented upon by Russian officials, they were often explicitly represented as part of the terrorist threat – as in the following statement by head of the Russian VVS (Airforce): 'In the objects that are attacked there should not be civilians, but if there are, it means that they have some kind of connection to the terrorists'.[31] Indeed, Russian military officials justified the bombing of the entire Chechen territory with reference to some specific oil installation or village somehow being connected to the terrorists.[32] The bombing of a Red Cross-marked civilian convoy on 29 October 1999 was, according to Russian officials, carried out against vehicles carrying Chechen fighters.[33]

Other times civilian casualties were directly blamed on the other side. Putin, for example, stated in an interview on radio Ekho Moskvy that the fighters 'are themselves shooting the peaceful population, who want to cooperate with the Federal Forces'.[34] The spate of official statements on the infamous bombing of the Grozny Central Market on 21 October either denied any involvement of the Russian armed forces and blamed the violence on the 'bandits' or 'terrorists', or admitted Russian involvement but justified it by representing the casualties/ targets as being part of the terrorist threat.[35]

Even years later, Russian official language offered representations of Chechnya as an existential terrorist threat as the main rationale for massive and indiscriminate bombing. The verdicts passed by the European Court of Human Rights in Strasbourg on cases against Russia in connection with the Second Chechen War include references in defence of the Russian government. The most common justification for unleashing heavy airpower, even against civilians, was that the use of lethal force was 'absolutely necessary' and 'proportionate' against the magnitude and violence of the 'illegal armed formations', as the Chechen fighters often are called in these documents. Also recurrent is the claim that the fighters were using civilians as 'human shields' or sabotaging efforts by the Federal Forces to secure safe exit for civilians.[36] In short, the massive violence employed by Russia is represented as 'normal' and reasonable given the nature of the threat. Russia was not to blame.

These statements by Russian officials on controversial bombing incidents in Chechnya show how the enactment of such practices was accompanied by linguistic representations that were linked to and referred to the dominant discourse on Chechnya as an existential terrorist threat. Total denial of Russian guilt, blaming civilian casualties on the 'terrorists' and representing civilians as part of the terrorist threat – all this is fully in line with the construction of Russia as innocent and of the Chechen threat as inhuman, capable of gruesome deeds and overwhelming in magnitude. Thus, Russian official language during (and after) the war drew on the core securitizing narrative on Chechnya. It justified the practices undertaken, even when 'shocking' results of this practice became evident, as with the loss of civilian lives. In turn, the re-iteration of these linguistic representations as well as their material enactment served to uphold and strengthen the discourse on Chechnya as an existential terrorist threat over time.

This is not to say that the dramatic results of war could not have created ruptures in the hegemony of the discourse on Chechnya as an existential terrorist threat. Alternative representations of these events (the bombing of Elistanzhi, the Grozny market, the Red Cross-marked civilian convoy, Samashki, Novy Sharoy) could have been launched, assigning the identity of victim to Chechnya/ Chechens and that of an existential threat to Chechen civilians to the Russian forces. And, indeed, a search in the database Pulic.ru for statements by the Russian political elite, experts and journalists on these events revealed that alternative positions were articulated, but only in marginal publications and voiced by a few.[37] Following the bombing of the Grozny market on 21 October, a few voices in the expert community were raised.[38] There was even one journalist account, an 'on the scene' report documenting and detailing the event by Andrey Babitsky and Maria Eismont, directly dismissing the official version.[39]

Amongst the political elite in the Federal Assembly, however, the potentially shocking results of massive bombing were not followed by any new and alternative articulations of Chechnya and Russia. My search on Public.ru did not reveal a single statement contradicting the official narrative of these events by members of the Duma or Federation Council, although there were several that confirmed it. For instance, the Chair of the Security Committee in the Duma Viktor Ilyukhin termed the explosions in the Grozny market on 21 October 'a huge provocation against the Russian armed forces, and against the military campaign aiming to frame the Federals as barbarians'. He went on to say that it was a provocation against Putin, aimed at discrediting Russia in the eyes of international public opinion.[40] In response to the emerging internal criticism of Russian warfare in Chechnya, Duma Member Viktor Chernomyrdin said:

I categorically condemn those of Russia's internal forces who conform to anti-Russian Western circles, dramatize the hysteria around the 'humanitarian catastrophe', and call for a halt to military operation and starting the negotiations.... Negotiations are not carried out with bandits. Bandits are killed for those who want to live and work normally.[41]

The big, mainstream newspapers maintained their one-sided discourse on Chechnya as an existential terrorist threat. In a similar fashion as official statements, events causing civilian casualties drowned in the stream of news on the Chechen 'terrorists', 'extremists' or 'bandits' and their terrible deeds, as well as Russian military activity to counter this threat all over Chechnya. During the intense bombing of Chechen targets in late September and in October, news reports in *NeGa* focused on technical descriptions of the military attacks, with hardly a word or picture presenting the victims of these bombardments.[42] 'Elistanzhi', 'Samashki', 'Novy Sharoy', places that in human rights reports are associated with indiscriminate bombing and heavy civilian casualties, were simply legitimate targets in these accounts.[43] Civilian casualties resulting from Russian bombing, such as the bombing of the Red Cross-marked civilian convoy on 29 October, were presented as highly unlikely.[44] A rare incident when the killing of several Spetsnaz soldiers was reported as collateral damage in some news outlets in early October was countered by *RoGa* reporting 'the truth'. Referring to the investigation of the bodies by experts, *RoGa* noted that: 'Fingers had been cut off. They were perforated with bursts from automatic weapons. The bandits obviously killed off the wounded soldiers: such wounds are not made by rockets, bombs or grenades.'[45]

Only one article in *NeGa* during autumn 1999 depicted a suffering Chechen civilian population and ascribed the guilt to *both* sides, noting that 'this war distinguishes itself from the former war by the particular cruelty of both sides'.[46] Otherwise, there was no mention of violence committed by the Federal army against Chechens, although several accounts detailed Russian humanitarian help.[47]

During the heavy bombardment of Grozny, *NeGa* journalists assured their readers that 'the Russian military did not stop repeating that they had recommended that the civilian population should leave the city. The corridor for refugees was open 24 hours a day'. The representation of the Russian military was contrasted to the Chechen fighters who 'use anything to protect themselves; in this way they are turning civilian objects into military objects'.[48] Very similar representations were given in *RoGa* articles during these days of October. Even when heavy bombardment was reported as carried out by federal air forces there was no mention of casualties among the civilian population: on the contrary, it was noted how 'the bandits mine the houses of ordinary Chechens and explode them whenever federal helicopters or planes appear in the sky. All this is done to set the civilian population against the federal power'. At the same time, it was noted how, thanks to the Russian authorities, 'pensions were paid out for the first time in three years and wages for doctors, teachers and administrative workers'.[49]

As regards the bombing of the Grozny market, the official versions of events and the accompanying representations were not questioned, but were simply reproduced in most Russian newspapers.[50]

Such reproduction of official discourse was particularly evident in reporting on the evolving refugee situation. Instead of a report on how the tens of thousands of Chechen refugees arriving in Ingushetiya were faring, the *NeGa* front-page article on October 1 was titled 'There is no humanitarian catastrophe in Ingushetiya yet – Some forces are prepared to use the refugees as a pretext to blame Russia'. The article went on to say that 'certain media outlets (together with human rights defenders) are participating in an information war against Russia, launched by Western security services – to undermine the policies of the Russian power against the terrorists and the band formations'. That Chechen refugees were prevented from crossing the border into Stavropol krai because the border was totally sealed off was 'completely understandable' because of 'the large number of other internal refugees in this krai, because of the terrorist acts and because of the repeated raids into this region from the Chechen side of the border'.[51] In other newspapers, the practice of bombing Chechen territory and simultaneously closing the border was justified with even more explicit references to the nature of the Chechen threat: 'There is no guarantee that there will not be a flow of under-aged kamikazes out of Chechnya ready to prepare terror acts against Russia from the refugee camps.'[52]

When thousands of fleeing Chechens were holed up outside the Kavkaz checkpoint on 29 October because the border was not opened to let them out as promised that day, that was reported with a small note in *NeGa*.[53] Even when more descriptive accounts of how the Chechen refugees were faring in Ingushetiya did appear, they were *not* accompanied by pictures of crisis and chaos. The key message was that even if the Russian authorities were responsible for the lack of humanitarian aid, and there had been civilian casualties, the refugees were not blaming the Russian authorities: 'this time around everybody is afraid of them [the Chechen fighters] – the soldiers, the journalists and even the Chechen refugees'.[54] Taken together, these reports on the Chechen refugees did not depict and detail them as victims or fellow human beings, but merely as a faceless and insignificant outcome of the Chechen threat.

Representations of Chechens as victims and fellow human beings were found only in small news items referring to statements by Russian human rights campaigners under headings such as 'Human rights defenders announce a "humanitarian crisis" in Chechnya'. For instance:

> the peaceful population in Chechnya is fleeing to save their lives from the shooting and bombs of the Federal forces and the threat of *zachistki* ... there is a catastrophic shortage of food and medication.... The refugees from Chechnya are trapped. They cannot return home as their houses are destroyed. They cannot move on into Russia from Ingushetiya. The Commander of the 'Zapad' group of federal forces General Shamanov has ordered that all roads out of Chechnya to Ingushetiya should be closed....

The federal forces during this 'counterterrorist operation' are killing peaceful inhabitants ... 2,000 people have been killed, many of them women and children.[55]

This is *not* the language of the journalists: they were directly quoting the human rights defenders. In the wider setting of *NeGa* reporting, this discourse on Chechnya was totally alien and marginal.

Instead, the Chechen side was represented as responsible for the plight of civilians during the Russian bombings of Chechnya. For example, the flow of refugees into Ingushetiya, fleeing the bombing, was presented as being provoked by the Maskhadov leadership to make the situation look like a humanitarian catastrophe and thus unleash Western criticism of Russia. The civilian deaths in the village of Elistanzhi were presented as the result of a massacre committed by Basayev.[56]

Conclusion

Why then was the official version of events credible to the Russian audience: why was it not contested? We cannot disregard the restrictions on the press and the lack of information. Yet, the answer lies as much in the categories that had been created as in increasing media control. Given the now-ingrained understanding of Russia as innocent and Chechnya/Chechens as guilty and dangerous, terrible deeds were most logically pinned on the Chechen side. Bestiality such as using humans as shields, for example, was only to be 'expected' from Chechen fighters.

At the outset, massive bombing of Chechnya did not appear unreasonable, against the background of an official discourse which reduced Chechnya to 'a huge terrorist camp' (see Chapter 5). Massive bombing was a logical enactment of policy statements referring to 'the toughest possible measures', 'hard', 'decisive', 'energetic' and 'uncompromising'. Even the targeting of civilian Chechens was to some extent foregrounded in linguistic representations that failed to delineate 'Chechens' from 'terrorists'. The use of massive violence may be an 'old' practice – but this particular instance of intensive bombing became acceptable through the congruence between level of threat implied in linguistic representations of Chechnya and the violence undertaken against Chechnya.

Tracing developments over time has shown constant references to and reiterations of representations in the 1999 securitizing narrative, justifying the bombing of Chechnya, particularly with potentially 'shocking events'. Such events were not followed by any major changes in representations of Chechnya and Russia. The 'discourse of reconciliation' did not re-emerge in the Russian public debate, as might have been expected. Instead, the statements which framed these 'shocking events' served to reinforce the dominant discourse on Chechnya as an existential terrorist threat – while Russia's identity was strengthened by references to its humanitarianism and innocence as the war progressed.

Finally, it is worth reflecting on how the continuous and heavy bombing of Chechen territory and targets in itself worked to confirm these very categorizations, along these lines: 'This threat must be extremely dangerous and omnipresent – just look at the way they have to bomb Chechnya'.

Notes

1 On 16 September, over 25,000 people gathered in Grozny to protest against the on-going Russian air strikes against dozens of towns and villages in southern Chechnya, Interfax reported. President Aslan Maskhadov said that over 200 people were killed in those raids, but Russian air force commander Anatolii Kornukov had told ITAR-TASS on 16 September that the raids were directed solely at guerrilla bases (*RFE/RL Newsline*, 20 September 1999).

2 *RFE/RL Newsline*, 20 September 1999.

3 'Chechnya snova lishilas' aviatsii', *Segodnya*, 24 September 1999.

4 'Nad vsey Checheney bezoblachnoye nebo', *Vremya*, 30 September 1999.

5 According to the same sources, more than 2000 fighters had been killed, and 250 support points destroyed as well as 150 terrorist bases and educational centres. ('Plany sukhoputnoy operatsii', *NeGa*, 28 September, 1999).

6 Marcel de Haas (2003) draws this conclusion in his study, *The Use of Russian Air-power in the Second Chechen War*. Swindon: Defence Academy of the United Kingdom, Conflict Studies Research Centre.

7 'Vokrug Groznogo szhimayetsya kol'tso', *NeGa*, 28 October 1999.

8 'Blokadnyy Grozny', *NeGa*, 29 October 1999.

9 'Boyevikam pred''yavlen ul'timatum', *NeGa*, 7 December 1999.

10 'Budet li zhdat' armiya?', *NeGa*, 8 December 1999.

11 'Defence dossier: Tactic Simply a War Crime', *Moscow Times*, 27 January 2000.

12 'Pobedy vysokaya tsena', *Ekspert*, 28 February 2000.

13 Marcel de Haas 2003, 15.

14 E. Gilligan. 2010. *Terror in Chechnya. Russia and the tragedy of civilians in war*. Princeton, NJ: Princeton University Press, 46.

15 'Rossiya ne poterpit na svoyey territorii nikakikh bandformirovaniy', *NeGa*, 7 December 1999.

16 No concerted national or international efforts have been made to calculate the number of casualties in the 1999–2000 bombing campaigns. According to estimates by human rights activists from Human Rights Watch, between 6,500 and 10,000 civilians died in the first nine months of the war (cited in A. Cherkasov and D. Grushkin. 2005. The Chechen wars and the struggle for human rights. In: R. Sakwa (ed.) *Chechnya: From past to future*. London: Anthem Press, 140).According to Gilligan (2010, 46), the lower number of civilian casualties during the second campaign compared to the first is related to the preparedness of the civilian population in 1999. In 1994 some 40,000 refugees crossed the border into Ingushetiya: in 1999–2000, as many as 250,000 did so.

17 The bombing of the Red Cross-marked civilian convoy on 29 October will be dis-cussed later. Another incident was the assault of Russian tanks on a bus full of refugees, killing 40, on 5 October (*NeGa*, 9 October 1999). See for example Memorial and Demos on the bombing of a civilian convoy near the village of Shaami Yurt (Memorial and Demos. 2007. *Counterterrorism Operation by the Russian Federation in the Northern Caucasus throughout 1999–2006*. Available at: www.memo.ru/hr/hotpoints/N-Caucas/dkeng.htm).

18 For example, the border at Sleptsovsk was closed on 23 October ('Obratnoy dorogi net', *Segodnya*, 30 October 1999). By 2 November, there were 20,000 refugees queuing to get out of Chechnya ('Chechen children shelled as they played', *Guardian*, 3 November 1999).

19 They are prohibited by the 1980 Geneva Convention.

20 Memorial and Civic Assistance. 1999. *The Report on the Observer Mission to the Zone of the Armed Conflict, Based on the Inspection Results in Ingushetia and Chechnya.* Available at www.memo.ru/eng/memhrc/texts/ch2.shtml and accessed 17 March 2016.

21 Memorial. 1999c. *The Missile Bombing of Grozny, October 21, 1999.* Available at www.memo.ru/eng/memhrc/texts/bom.shtml and accessed 17 March 2016.

22 As regards the operation in Chechnya in general, the Court in its decisions acknowledged that at that time the federal government would need to take exceptional measures 'in order to regain control over the Republic and to suppress the illegal armed insurgency. These measures could presumably include employment of military aviation equipped with heavy combat weapons' (Isayeva v. Russia par. 178). (The ECHR decisions are available at www.srji.org/en/legal/cases. To find the summary of the case plus a link to the full text, put the name of the applicant into the 'text search').

23 See, for example, also Isayeva v. Russia, Mezhidov v. Russia, Abuyeva and others v. Russia. (Available at www.srji.org/en/legal/cases. To find the summary of the case plus a link to the full text, put the name of the applicant into the 'text search').

24 See for example UK and U.S. reactions in 'UK condemns Chechnya ultimatum', *BBC News*, 7 December 1999 and 'Russia will pay for Chechnya', *BBC News*, 7 December 1999.

25 J.F. Baddeley. 1908. *The Russian Conquest of the Caucasus.* London: Longmans, Green.

26 'Russia blamed for attack on refugee bus', *Guardian*, 8 October 1999.

27 Referred to in 'Chechenskiy uzel opornyye punkty boyevikov-v ogne', *Yakutiya*, 2 November 1999.

28 On 7 October 1999, two Russian Sukhoi Su-24 fighter bombers dropped several cluster bombs on the apparently undefended mountain village of Elistanzhi. At least 34 people were killed (48 according to some reports) and some 20 to over 100 people in the small village were wounded, mostly women and children. At least nine children were reportedly killed when one bomb hit the local school. (*Voice of America* report, 7 October 1999, available at www.globalsecurity.org/military/library/news/1999/10/991009-chechen1.htm, and accessed 17 March 2016.) Memorial and Civic Assistance visited Elistanzhi shortly after the bombing and found no evidence of any Chechen separatist military presence in the village (Memorial and Civic Assistance 1999).

29 Cited in 'Gde zhe Basayev?', *NeGa*, 13 October 1999.

30 Press Agency of the Russian Ministry of Defence, cited in 'Chechnya', *Sankt-Petersburgskiye Vedomosti*, 29 October 2008.

31 'Plany sukhoputnoy operatsii', *NeGa*, 28 September 1999.

32 General Vladimir Shamanov, cited in 'Rossiya ne dolzhna opravdyvat'sya za svoye stremleniye pokonchit' s terrorizmom', *NeGa*, 4 November 1999 or for example 'Khronika konflikta', *NeGa*, 13 November 1999.

33 'MKKK podtverzhdayet', *Federal'noye Agentstvo Novostey*, 31 October 1999.

34 Cited in 'Chechenskiy uzel opornye punkty boyevikov-v ogne', *Yakutiya*, 2 November 1999.

35 A Memorial report (1999d) collected official statements following the event, 'On 22 October, the RF authorities of different positions gave at least five essentially different comments on the event the day before'.

36 See, for example, 'Case of Isayeva, Yusupova and Bazayeva v. Russia, Judgement, Strasbourg, 24 February 2005', 'Case of Isayeva v. Russia, Judgement Strasbourg, 24 February 2005' (Available at www.srji.org/en/legal/cases. To find the summary of the case plus a link to the full text, put the name of the applicant into the 'text search').

37 This alternative version could sometimes be found in certain smaller newspapers ('Terek pereyden', *Ekspress-Khronika*, 25 October 1999, 'Grozny resheno sdelat'

tikhim', *Novaya Gazeta*, 18 October 1999, 'Lychshe by atomnoy bomboy', *Sobesednik*, 25 November 1999); at times in the regional press ('"Federaly voyuyut protiv mirnogo neseleniya',-utverzhdayet ochevidets Chechenskoy voyny', *Khronometr*, (Ivanovo), 24 November 1999, 'Khronika pikiruyushchego shturmovika, *Monitor*, (Nizhny Novgorod), 13 October 1999) and in statements by 'traditional' human rights advocates such as Sergey Kovalev ('Sergey Kovalev: Dve voyny', *Novoye Vremya*, 28 November 1999) or Memorial ('Pravozashchitnyy tsentr Memorial utverzhdayet, chto v Chechne primenyayutsya kassetnyye bomby', *Federal'noye Agentstvo Novostey*, 26 October 1999).

38 Dmitry Trenin in 'Slabost' geopoliticheskogo myshleniya', *Nezavisimoye Voyennoye Obozreniye*, 19 November 1999; and Andrey Piontkovsky in 'Chtoby pobedit' v etoy voyne, nado ubit' ikh vsekh', *Novaya Gazeta*, 1 November 1999.

39 'Bombili', *Vremya, MN* 26 October 1999.

40 'Psikhologicheskaya voyna v Chechne razgorayetsya', *NeGa*, 26 October 1999.

41 'My razberemsya s Chechney bez pomoshchi NATO', *Argumenty i Fakty*, 8 December 1999.

42 'Voyna bez vykhodnykh', *NeGa*, 25 September 1999; 'Plany sukhoputnoy operatsii', *NeGa*, 28 September 1999; 'Terroristicheski front na vode', *NeGa*, 8 October 1999; 'Neuzheli eto Pravda?', front page *NeGa*, 14 October 1999; 'Vokrug Groznogo szhimayetsya kol'tso', *NeGa*, 28 October 1999; 'Novaya granitsa Ichkerii', *NeGa*, 6 October 1999. There were some exceptions, such as 'V Prigranichnykh s Chechney rayonakh rastet chislo bezhentsev', *NeGa*, 6 October 1999.

43 'Obstanovka v Severo-Kavkazskom regione na 25 Oktyabrya', *Krasnaya Zvezda*, 26 October 1999; 'Artilleriya Federalov prodolzhayet massirovannyy obstrel pozitsii boyevikov', *Federal'noye Agentstvo Novostey*, 12 November 1999; 'Chechnya: Khronika konflikta', *Nezavisimoye Voyennoye Obozreniye*, 5 November 1999; 'Chechnya: Khronika konflika', *NeGa*, 30 October 1999; 'Obstanovka v Severo-Kavkazskom regione na 12 Noyabrya', *Krasnaya Zvezda*, 13 November 1999.

44 'V Chechne bol'she shansov stat' zalozhnikom, chem 'grusom 200', *Novyye Izvestiya*, 1 November 1999.

45 'Zachem putat' boyevikov so shturmovikami', *RoGa*, 15 October 1999.

46 'Blokadnyy Grozny', *NeGa*, 29 October 1999.

47 'Despite the continuing struggle against the fighters, life in the liberated areas (of Chechnya) is returning to normal. For the past week, the humanitarian trucks have been arriving from Russia ... they have started to pay out pensions' ('Federal'nyye voyska podoshli k Groznomu', *NeGa*, 21 October 1999). For similar account, see 'Psikhologicheskaya voyna v Chechne razgorayetsya', *NeGa*, 26 October 1999 and 'Vokrug Groznogo szhimayetsya kol'tso', *NeGa*, 28 October 1999.

48 'Grozny budut brat' po chastyam', *NeGa*, 22 October 1999, also 'Vokrug Groznogo szhimayetsya kol'tso', *NeGa*, 28 October 1999.

49 'Kavkazskiy uzel: 18 Oktober', *RoGa*, 19 October 1999.

50 See, for example, 'Vzryvnaya volna vernulas' v Chechnyu', *Segodnya*,23 October 1999 or 'Chernyy rynok v Groznom byl unichtozhen svintsovym udarom 'Grada'', *Komsomol'skaya Pravda*, 23 October 1999.

51 *NeGa*, 1 October 1999.

52 'Obratnoy dorogi net', *Segodnya*, 30 October 1999.

53 'Granitsa ostalas' na zamke', *NeGa*, 29 October 1999.

54 'Bezhentsy khotyat mira i khleba', *NeGa*, 20 October 1999.

55 'Pravozashchitniki zayavlyayut o 'gumannitarnoy katastrofe' v Chechne', *NeGa*, 19 October 1999.

56 'Chislo Zverya', *Versty*, 11 November 1999. See also, for example, 'Obstanovka v Severo-Kavkazskom regione na 28 Oktyabrya', *Krasnaya Zvezda*, 29 October 1999, or 'Obstanovka v Severo-Kavkazskom regione na 21 Oktyabrya', *Krasnaya Zvezda*, 22 October 1999.

12 Cleansing Chechnya

On 30 September 1999, the ground offensive into Chechnya started. There are many possible stories to tell and many alternative ways of presenting how the war was fought on the ground. There were many regular armed clashes and battles between Chechen and foreign fighters and Russian troops of varying stripes. There were ambushes at Russian garrisons and attacks with remote-controlled bombing devices. There were also atrocities committed by the Chechen and foreign fighters against the Chechen civilian population and against Russian soldiers. These events and many others are not included in the account that follows. Not because they did not happen or were insignificant, but simply because my concern is with how the seemingly unacceptable warfare practices undertaken by Russian forces in Chechnya during the Second Chechen War were enabled by Russian representations of Chechnya and Chechens. The account will therefore focus on the practices undertaken over the course of several years as part of the effort to 'cleanse' the entire territory of Chechnya of 'terrorists' in the *'zachistki'* (a slang word meaning 'cleansing operations') and the ensuing practices at 'filtration points' *(fil'tratsionnyy punkt).*

The treatment of people in the *zachistki* and at 'filtration points' was characterized by massive and arbitrary violence. According to Rachel Denber of Human Rights Watch, human rights violations were carried out by Russian troops on a much wider scale during the second campaign in Chechnya than the first.[1] These were emergency measures that went far beyond both legal provisions in armed combat and what one would think was socially acceptable to the Russian public.[2] And indeed, as we shall see below, reports of gross human rights violations in Chechnya in connection with these practices of war were proceeded by a rupture in the discourse which constructed Chechnya as an existential threat and Russia as the righteous defender – but only to a limited degree. Generally, these practices did not seem to be wholly unacceptable or illegitimate any longer. Again, the argument I will be advancing is that the system of *zachistka* and 'filtration' as well as the blunt violence used in connection with these practices now appeared both logical and legitimate because it matched identity constructions of Chechnya and Chechens found in the official securitizing narrative as well as in representations voiced by the political elite, the experts and the journalists.

This chapter looks at Russian practices of war in connection with the ground offensive in Chechnya (from 30 September 1999 onward). It presents the practices of *zachistka* and 'filtration' employed in Chechnya in 2000 to 2002 in order to 'cleanse' the territory of terrorists, and argues that these practices acquired a systematic character during the Second Chechen War. It then moves on to discuss the co-existence of these practices of war with language, suggesting that they were mutually constitutive. It examines how the securitizing narrative prepared the ground for these practices and investigates how various aspects of official representations were echoed in the language of generals and soldiers as these practices were undertaken. Finally, it looks at how 'shocking' revelations of atrocities against civilians were justified in official statements with reference to the initial securitizing narrative, and discusses why a broad public opinion against the war in Chechnya did not emerge.

Cleaning up and filtering out 'terrorists'

The first *zachistka* in the Second Chechen War (in Borozdinovsky in the Shelkovskiy region) was noted by Russian newspapers in early October 1999.[3] As early as the end of November, more than 80 'populated points' (*naselënnyy punkt'*) had been subjected to this procedure, according to official sources.[4] The deal allegedly offered the civilian population before the *zachistka* was that troops would not enter the village if they were allowed to check whether the fighters had left in such a 'cleansing' operation.[5] This was in line with the definition of a *zachistka* often given by Russian officials as 'a special operation aimed at checking people's residence permits and identifying participants of illegal armed formations'.[6] Apart from this rationale, there were few legal instruments regulating the *zachistka*. The Law on the Suppression of Terrorism (1998) provided wide-ranging powers to those conducting a counter-terrorist operation[7] and the *zachistki* were undertaken without interference from public prosecutors. The treatment of detainees at 'filtration points' (discussed below) was never subject to normal due process during the Second Chechen War either.[8] Human rights organizations have claimed that military, police, and security service units conducting such operations in Chechnya routinely interpreted the silence of the anti-terrorism law as regards procedural matters to mean that no standards of due process should be followed.[9] What weight, then, does the securitizing narrative acquire as an instruction on 'what to do' in such a legal and procedural vacuum? Despite the difficulties of recording abuses and atrocities during the war, there can be no doubt that the hallmark of the *zachistka* became arbitrary detainments, torture, rape, looting, killing and the 'disappearance' of civilians.

The *zachistki* conducted in Alkhan-Yurt in December 1999, in the Staropromyslovsky district of Grozny in January 2000, and in Novye Aldy in February 2000 were widely covered also in the Russian media. In the two weeks following 1 December, Russian forces, according to Human Rights Watch (HRW), 'went on a rampage' in the village of Alkhan-Yurt south of Grozny, systematically looting and burning down the village, summarily executing at least 14 civilians as well as conducting rapes and torture.[10] In the *zachistka* of Staropromyslovsky district in Grozny between late December and mid-January, Russian soldiers summarily executed at least 38 civilians, according to testimony taken by HRW in February 2000. HRW reported that the victims were women and elderly men, and that they appeared to have been deliberately shot by Russian soldiers at close range. Russian soldiers also committed many other abuses in the district, including looting and destroying civilian property and forcing residents to risk sniper fire to recover the bodies of fallen Russian soldiers. Six men who were last seen in Russian custody 'disappeared' from Staropromyslovsky during this same period.[11] In Novye Aldy, Memorial reported, 56 innocent civilians, including old men, women and even a one-year-old baby, were summarily shot down on 5 February.[12] Houses and dead bodies were set on fire; Russian contract soldiers returned several times afterwards to loot in Novye Aldy.[13]

When large-scale battle was replaced by guerrilla warfare in the summer of 2000, *zachistki* became more frequent. Gilligan collected information and details on the abuses and torture methods used during *zachistki* of the villages of Shuani in July, Gekhi in August and Chernorechye in late August/early September 2000. During winter 2000–2001 there were *zachistki* in Grozny, Kurchaloy, Mayrtup, Chernorechye, Chiri-Yurt, Tsotsin-Yurt, Novye Atagi, Argun and several times in Alkhan-Kala and Starye-Atagi.[14] Large-scale *zachistki* took place in summer 2001 in Alkhan-Kala from 19 to 25 June, in Sernovodsk on 2 and 3 July and in Assinovskaya on 3 and 4 July.[15] The two last-mentioned were widely covered in the national and international press because of the location of these villages close to the Ingush border. Memorial has also documented the repeated *zachistki* of Tsotsin-Yurt in 2001 and 2002, numbering 40 in all.[16] According to Memorial, the practice of *zachistka* was widely used until November 2002 when the Russian President declared that broad-scale operations should not be held in Chechen towns and villages. Thereafter the number of large-scale *zachistki* went gradually down, decreasing sharply after summer 2003.[17]

Gilligan's accounts[18] as well as that of Memorial and various other reports indicate that the pattern of public executions was replaced by *disappearances* from summer 2000 onward, but that degrading treatment, violence, torture, extra-judicial killing and robbery remained continuing and routine features of *zachistki* throughout this period.[19]

The actual extent of killings and violence against civilians in connection with these *zachistki* is difficult to determine. I have cited documentation by human rights organizations in connection with the best-known *zachistki*. Several of their findings have been confirmed since then in ECHR decisions that establish the existence of extrajudicial killings, disappearances and torture in these operations.[20]

Exact and correct figures are difficult or even impossible to establish however, they are not of primary interest for this study. Suffice it to say that the *zachistka*, with all the violence that accompanied such an operation, was a key practice of war/emergency measure employed in Russia's fight against the Chechen 'terrorist' threat. While *zachistki* also took place during the First Chechen War, they were practised on a larger scale and in a systematic fashion during the Second Chechen War. That the *zachistki* had by 2003 acquired status as a 'normal' or 'routine' practice for disciplining Chechnya/Chechens was evident during the preparations for the referendum on a new Chechen constitution initiated by the Russian leadership in spring 2003. Villages that refused to vote in the referendum in March 2003 were threatened with *zachistki*.[21]

The system of *zachistki* was paired with a system of 'filtration points', a broad label for detainment facilities, whether legally based temporary detention facilities and pre-trial detention facilities or places with no official status (such as a field outside a village, a pit in the ground, military vehicles, tents, abandoned buildings or a military commander's office). The rationale behind the 'filtration points' was to identify and filter out participants and supporters of the armed resistance in Chechnya, as well as creating a network of informants among the local population. That civilians would be taken out for some kind of 'filtration' is obvious in a counter-insurgency war. Fighting such a war without undertaking practices aimed at distinguishing fighters from civilians is difficult.

Nevertheless, if there is one feature that is striking about detentions in Chechnya over the years it is how *arbitrary* they were. In numerous witness accounts in the human rights reports referred in this chapter, men were simply 'taken', without any check of their identity.[22] As Holly Cartner, executive director of the Europe and Central Asia division of HRW, noted in February 2000: 'in many of these cases the arrest appears to be based solely on the ethnic background of the men'.[23] According to Memorial, the major characteristic of the 'filtration system' was its non-selectivity.[24] The ever-increasing number of men, women and even children checked and 'filtered' in this system seemed to go beyond the rationale of finding fighters. It sometimes looked more like the targeting of an entire group of people, particularly when this non-selectivity was combined with inhumane treatment and excessive violence during detention at 'filtration points'.

As mass non-selective detentions of local residents became a distinct feature of the *zachistki* from 2000 onward, they, together with detentions at border crossings or 'checkpoints' *(blokposty)* established across Chechnya to restrict the movement of men and boys within the republic, created a flow of people into various 'filtration points'. Such 'temporary filtration points' were usually established on the outskirts of villages and towns. While many of the detained were released after being checked, practically all those held at 'temporary filtration points' were exposed to beatings and torture, according to Memorial.[25] The torture methods included beating and kicking people until they could no longer stand on their feet, putting plastic bags over the heads of detainees causing asphyxiation, mock executions, rape, placing detainees in painful positions, forcing them to stand outdoors in the cold without clothes, or lying face down in the heat for hours. There were numerous reports of detainees (also children) given electric shocks (to the genitals, toes and fingers) or being cut with knives or tear-gassed.[26] Sometimes detentions resulted in the death or execution of detainees. Bodies were found with gunshots to the head, ears cut off or without heads; sometimes the bodies were so grossly disfigured that identification was impossible. Sometimes the bodies were found dumped in a well, by the roadside or in makeshift graves.[27] 'Disappearances' became an increasing feature from summer 2000 onward, often following in the wake of detainments at 'temporary filtration points'.[28]

Those not released from 'temporary filtration points' were transferred to more long-term official or unofficial detention facilities, or to illegal prisons. The most notorious of such facilities, located at Chernokozovo, acquired official status as a pre-trial detention centre (more precisely, an 'investigative isolator' (SIZO) subordinate to the Ministry of Justice) only after grave abuses were exposed during winter 2000. As early as January, reports emerged on the practices employed at Chernokozovo, where hundreds of men but also women and children were detained. According to testimonies of several survivors, detainees were welcomed with the words 'Welcome to Hell' and forced to walk through a human corridor of men armed with clubs and hammers ('the gauntlet'). Detainees were stripped of warm clothes and valuables. Some were kept in pits, others in cells without toilet facilities. Sometimes detainees were ordered to stand with their hands raised for entire days. They were regularly taken out for interrogation and tortured, particularly at night. During interrogation, detainees were exposed to electro-shocks, systematic and severe beatings, also genital beatings, and teeth were sometimes sawn off. There has also been convincing testimony of rape and sexual assault at Chernokozovo.[29]

While Chernokozovo is the best-known camp, widely exposed in national and international press during the winter of 2000, there existed other, more permanent detention facilities as well, both inside and outside Chechnya. Documenting the numbers and whereabouts of such facilities during the Second Chechen War was difficult because of extremely restricted access for journalists and human rights organizations, and the lack of official openness on the counter-terrorist campaign in Chechnya. On 24 March 2000, Amnesty International indicated that such facilities existed in the villages of Kadi-Yurt, Urus-Martan, Tolstoy-Yurt, Chiri-Yurt as well as in Grozny.[30] Memorial confirmed some of these allegations and noted that, during the period 2000 to 2002, the SIZO functioning under the Ministry of Interior Departments in Urus-Martan and the Oktabrskiy district of the city of Grozny became especially notorious. The detained and arrested persons there were regularly exposed to torture; some detainees 'disappeared'.[31] Memorial also identified a long-term detention facility named 'Titanic' by the military, located between the villages of Alleroy and Tsentoroy.[32] Other detention facilities were reported to exist in the towns of Mozdok and Grigoryevsk in the Stavropol region. There were pre-trial detention centres (SIZO) in the town of Pyatigorsk and in the city of Stavropol, in the Stavropol Region. Judging by the testimonies collected from these places, the use of torture and inhumane treatment appears to have been widespread and systematic. Certain practices, such as forcing detainees to run 'the gauntlet', were commonplace.[33]

As part of the filtration system there was also a set of illegal prisons. Some of these were established close to military bases or places where special units of the Ministry of Interior were deployed.[34] The most notorious such illegal prison was at the military base in the village of Khankala. According to Memorial, prisoners at Khankala were not officially registered anywhere, neither as detained nor as arrested. Most of them were held in holes dug in the ground, or in trucks and railway cars intended for prisoner transport.[35] That a similar pattern of abuse and violence at detention facilities as that noted above accompanied the treatment of detainees at Khankala became evident when a grave of 51 bodies was discovered in the village of Dachny Poselok close to Khankala in February 2001.[36] There was also a facility near the military base at Mozdok as well as at other military encampments. Human rights organizations have documented the systematic use of torture at these facilities also.[37]

Yet again the point of this account has not been to give a full overview of the various 'filtration points' in Chechnya, but to substantiate the claim that the practice of arbitrarily detaining and subjecting people to violence acquired a systematic and mass character during the Second Chechen War. As with the *zachistki*, the 'filtration points' were nothing new. They, as well as the term 'filtration point', had appeared during the First Chechen War, but at the time their use was unofficial and controversial. During the Second War, 'filtration points' became legitimized and institutionalized when some of the filtration system facilities (such as at Chernokozovo) got the status of investigative isolators (SIZO) under the Ministry of Justice and temporary detention isolators (IVS) under the Ministry of Interior.[38] Moreover, even if we again are left without exact figures to compare, there is no doubt that the 'filtration' of the Chechen population was massive and systematic. To cite Memorial:

> The exact number of the people having passed through the filtration system is impossible to identify – those are thousands of citizens.... Thus, by the most modest estimations, the overall number of those having passed through the 'filtration system' reaches 200,000. For Chechnya, with its population at present being less than one million, it is an enormous number.[39]

How language matters II

The puzzle I am seeking to address is not the motivations behind such brutal and systematic violence against what were often innocent civilians, but how such violence was made possible by the categories and distinctions that had been created in the securitizing narrative. To start at the basic level, the attempts to 'cleanse' and 'filtrate' practically the entire Chechen territory and the entire population were not without logic, as Chechnya was represented in official discourse as 'a huge terrorist camp' and the terrorist threat as elusive, yet powerful and omnipresent. Blunt and indiscriminate violence against those living in Chechnya was also logical, given the equation of Chechnya with the terrorist threat and the extremely de-humanized, cruel and dangerous nature ascribed to the 'terrorists' in the official securitizing narrative. Russian forces were not fighting fellow human beings.

Moreover, the practices undertaken in Chechnya were in many ways a direct reflection of the 'way out'/emergency measures featured in official language. In Chapter 5, I argued that official statements in 1999 dismissed the use of law or understanding as a means of dealing with the threat. There should be 'no forbearance with bandits'.[40] Thus, the arbitrary, illegal and merciless violence used in connection with *zachistki* and at 'filtration points' was not outside the bounds of 'policy advice' given in official rhetoric.

On the contrary, 'the toughest measures possible' had been the key instruction on how to conduct the counter-terrorist operation[41] along with other words that pointed in the same direction (such as 'hard', 'tough', 'decisive', 'energetic' or 'uncompromising', the terrorists needed to be 'annihilated' and 'destroyed', any kind of soft approach would mean the destruction of Russia).[42] The stories of violence and brutality in testimonies from Chechnya are in many ways a logical enactment of the 'way out' indicated in the official securitizing narrative.

I also want to suggest that the securitizing narrative launched by the Russian leadership at the outset of the campaign which was emphasized, confirmed and broadened in accounts by the political elite, experts and journalists found a parallel in statements by Russia's top military service men. One highly placed MoD officer described the juxtaposition between Chechen violence and lawlessness and Russian law, order and normality in this way:

Only the land where our soldiers are stationed can be considered ours. Otherwise it is enemy territory.... It has been four days since we left for the Terek, and only today did we begin cleaning up Borozdinovsky. All administrative functions have to be executed by the military. The Chechens have been demonstrating their 'ability' for governance for ten years now. They didn't manage to live by Russian, *Sharia* or even thieves' rules. It is quite possible that most of them do not want to fight with Russia. However, after years of lawlessness, stealing petroleum products, dealing drugs and keeping slaves has become a habit for very many. An entire generation has grown up that cannot do, and knows not of anything but, murder and theft. In order to once again bring all these people under the rule of the law, it is necessary to spend a lot of effort, demonstrate firmness, steadfastness, but at the same time flexibility. Otherwise all the effort and sacrifice will be for nothing. You can destroy all of the military commanders, but an 'appeased' but not decriminalized Chechnya will breed new Basayevs and Khattabs over and over again. The Chechens will not be able to deal with the task of returning to normal life on their own.[43]

Detailed comparison of the various components of the official securitizing narrative with statements by Russia's top military servicemen reveals that this narrative reverberated in the language of those who led the military campaign and thus served to legitimize the brutal practices as they were undertaken. When the first phase of the 'counter-terrorist operation' was declared over, and Russian forces controlled one third of Chechen territory in mid-October, General Kulakov announced that the second phase would focus on 'rooting out the terrorists in the entire territory of Chechnya'.[44] As the war proceeded, several well-known features from the Russian leadership's narrative were echoed in the language of the country's top military and security personnel. Representations of the adversary as 'terrorist' or 'extremist' were recurrent features of their reports from Chechnya.[45] So was the concept of the close bonds between Chechnya and international terrorism. One Russian general even indicated that Arabs made up the core of the fighters in Chechnya.[46]

The cruel nature of the enemy was elevated in military discourse as well, often by referring to the fighters' illegal and extreme warring methods. Before taking Grozny, for example, the Defence Ministry announced that the Chechen fighters were planning to use mustard gas against Russian forces, as well as ammunition prohibited according to international conventions. Russian intelligence stated that there were still two thousand 'terrorists' left in the city and that they were planning to use all kinds of chemical weapons against the Russian forces.[47] Similarly, according to the military command and the security forces, reports that there were terrorists on their way out of Chechnya to commit terrorist acts in various regions of the Russian Federation and that they were hiding among the rows of refugees forced them to employ unprecedented measures to wipe out these groups and to employ the harshest possible measures at the checkpoints.[48] Special Forces soldiers serving in Chechnya were allegedly shown videos of Russian soldiers being tortured by rebels 'to make them feel vicious so that they would not feel any pity'.[49] This constant articulation of a 'lawless' and 'brutal' enemy opened up for and legitimized lawless and brutal practices on the part of Russian forces. The logic was captured in the words of a member of the Russian Special Forces serving in Chechnya: 'The only way to struggle with lawlessness is with lawless ways.'[50]

Representations that emphasized the cruel nature of the enemy combined with instructions given by Russia's top generals to 'annihilate', 'eradicate' or 'extinguish' this enemy indicated what were appropriate practices on the ground. During the offensive in Dagestan in August 1999, Army Chief of Staff Anatoly Kvashnin said: 'We are talking about the total annihilation of the militants.'[51] This 'total annihilation of the militants' required by Kvashnin found a parallel when soldiers entering Novye Aldy in February 2000 to carry out the *zachistka* cried 'get out, you sons of bitches, we'll kill you all, we have orders'.[52] Similarly, the language employed in connection with the establishment of a network of internal checkpoints/*blokposty* across Chechnya indicates what kind of treatment the detainees could receive. General Viktor Kazantsev stated:

> the measure is aimed at curbing the free moving of the militants under the guise of peaceful civilians.... Identity checks in liberated areas plus the

toughening of search procedures at checkpoints will put in very tough circumstances those who are inclined to call to arms and kill by night.[53]

My point is that such words by the generals open for a range of possible actions on the battlefield; brute and indiscriminate violence does not fall outside this range.

Moving to the frontline of the incipient battle, we find that representations among the Russian soldiers also seem to involve a distinct juxtaposition between the Russian Self and the Chechens as a dangerous and treacherous Other. On 15 October 1999, a Russian soldier contributed a piece in *RoGa* titled 'The territory beyond Terek is also ours', which describes the situation of Russian soldiers in battle. There is a thick description of the comradeship among the Russian soldiers, as opposed to 'the other side of the river where they had already placed price tags on their targets – for a pilot you could get 100,000 dollars, for an artillery commander 70,000'. The soldier concludes that:

> this is how they do it, they send the young ones first … they trick the Federal forces to direct the fire against innocent people in the villages in the night time. The next day they film the so-called bestiality of the Russian army and post it to all TV channels in the world … then the civilians want to avenge the death of their relatives, and this is how the bandits increase their ranks.[54]

As noted, brute and indiscriminate violence on the part of Russian soldiers had been widespread during the First Chechen War as well. The difference was that in 1999 the dense discourse on Chechnya as an existential terrorist threat made the execution of such violence on a mass scale both logical and legitimate. The illegal detention, torture or execution of a Chechen became all the more appropriate when this Chechen was construed as different and dangerous, not even human. As can be seen in the testimonies given to human rights organizations, even the statements accompanying the execution of these practices of war on the micro-level reproduced the representation of Chechnya/Chechens as an existential terrorist threat to Russia.

Questioning during detainment and torture often identified the detainees as 'Wahhabi', 'Arab', 'bandit' or 'fanatic'.[55] According to testimony by 'Sultan Eldarbiev', detainees at Chernokozovo were forced to sign confessions that they were 'Wahhabi'.[56] 'Akhmed Isaev' recounted how:

> they ordered me when I reached the door, to … say the words 'Citizen Officer, thank you for seeing me. I am [gives name]. According to your order I have crawled up here'. They also said that the faster I crawled, the fewer hits I would get. They laughed, saying I crawled like a 'Wahhabi'.[57]

Sultan Denoev told how:

> they put me against a wall, and said, 'in the name of the Russian Federation, according to Article 208 you will be shot.' This was in the second

interrogation. I said 'OK, my life is in your hands'. I just knew nothing would help. Then they got more angry and said, 'What, don't you want to live, are you a fanatic?'[58]

Such re-phrasing of the Chechen fighters as an international terrorist threat (in all its different aspects) certainly framed actions on the ground in Chechnya.

Other parts of the official securitizing narrative found a parallel in language on the ground as well. Of particular significance for understanding the legitimation of gross violence and degrading treatment were the collective references that served to de-humanize the victims. Testimonies in connection with extrajudicial killings show how the Chechens were also referred to as 'scum' by the soldiers.[59] Just as references to the 'terrorists' being 'animals' were frequent in official language,[60] detainees were frequently referred to as animals when they were subjected to torture and violence: 'You dogs, you sheep, you were killing our comrades. Now we will show you!'[61] During the massacre in Alkhan-Yurt one soldier allegedly shouted 'you animals, faggots, you should all be shot'.[62] Referring to a video that allegedly showed what Russian soldiers held hostage by the 'Wahhabis' had been subjected to, a soldier concluded 'Chechens.... They are not people. They have to be annihilated like rabid dogs'.[63]

Testimonies also reveal another pattern: illegal violence was frequently justified simply with reference to the victim being 'Chechen' and juxtaposing 'Chechen' against 'Russia'. When a Chechen detained after the *zachistki* of Assinovskaya and Sernovodsk was trying to convince his torturers that he was not guilty of anything, the soldiers answered that they couldn't care less – 'the main thing is that you are a Chechen'.[64] In connection with a *zachistka* of Tsotsin-Yurt and the ensuing detainment of more than 100 people at a 'temporary filtration point', Kazbek Khazmagomadov had asked why they were beating the detainees. The representatives of the federal forces answered 'because you are Chechens!'[65] Other victims reported how Russian soldiers told them they 'were bandits', who 'did nothing for the motherland',[66] or that they wished they could kill all Chechens: 'then Russia would be OK'.[67] The widespread beating of detainees on the genitals at checkpoints, in temporary filtration points and at more long-term detention facilities was also a practice enabled by the categorization of 'Chechens' as different and dangerous. Such beatings were carried out accompanied by words like 'you will never have children again'.[68]

Whether the motivation behind such language was feelings of racial hatred is irrelevant in this account. My point is that the merging of everything and everyone 'Chechen' into a category of 'existential threat' in layer upon layer of talk presented throughout this book, and the enormous distance such a discourse created between this group of people and 'Russia' made possible and logical the brutal treatment of Chechens on the battlefield. As these practices were carried out, they simultaneously confirmed the identity given to Chechens and Russia in linguistic structures. The dominance of 'Russia' over 'Chechnya' was also enacted and confirmed in these practices. Several testimonies from Chechnya recount how soldiers forced detainees to crawl and make them say things like 'Comrade Colonel, let me crawl to you', or 'request permission to crawl' and also wanted them to say 'thank you' for torturing and subjecting them to inhuman treatment.[69]

Shocking events change nothing

Finally, I want to broaden the focus, asking whether some of these potentially 'shocking events' on the battlefield in Chechnya were followed by a re-emergence of the 'discourse of reconciliation' in the Russian press, as well as how these events were represented in official statements, and to what effect.

Once again, it should be borne in mind that this war was not as visible to the Russian public as the First Chechen War had been. The changing media situation was vividly illustrated in this account by a villager in Novye Aldy, where 56 people were killed in one of the worst *zachiski*:

> We rigged up a motor to the television, and we watched central [Russian] television and heard that federal units had carried out a special operation to eliminate fighters in the village of Novye Aldy.... There were corpses lying not far from the television set – I'll never erase that picture from my mind.[70]

On the whole, a large number of *zachistki*, (sometimes referred to as a 'repeated *zachistka*', a 'soft *zachistka*' or a 'hard *zachistka*') were noted in various newspaper reports and chronicles on the counter-terrorist campaign – but were presented as regular, 'natural' and legitimate undertakings resulting in the confiscation of weapons and drugs and the detention of people or 'bandits' listed in the federal wanted list.[71] When the ground offensive in Chechnya was well under way, *NeGa* reports, for example, did not focus on pictures and details of suffering people. The campaign was presented in a matter-of-fact language as an orderly sequence of events without much human cost.[72] In these accounts, the names of villages or regions where human rights organizations had documented atrocities are simply noted, without any representation of violence and brutality.[73]

The representations of Russian forces being welcomed into Chechen villages as 'liberators' and creators of order, law, civilization and normality were extended into the period when atrocities by Russian forces were being committed. Russian newspapers offered no images or representations of Chechens as victims of Russian violence that could have altered the asymmetric power relation between Chechnya and Russia and made up the backbone of a re-emerging 'discourse of reconciliation'. An interview in a local newspaper with one OMON soldier from Khabarovsk returning from service in Chechnya in January 2000 illustrates how the representation of Russian–Chechen relations was perpetuated:

> in the regions liberated from the bandits our guys played the roles of peace builders, sharply distinguishing between extremists and people, as is usua

in the UN. Their presence aids the federal power in efforts to return
suffering people to normal life as quickly as possible.[74]

Nevertheless, news of the *zachistki* and the alleged atrocities committed by
Russian forces in connection with these operations in Alkhan-Yurt,
Staropromyslovsky and Novye Aldy did make it onto the pages of Russian
media. The reception in Russian journalistic accounts was one of disbelief.
Indeed, several papers represented or dismissed reports on atrocities as anti-
Russian Western or Chechen propaganda.[75] Some did not even report on alleged
atrocities by Russian soldiers at all, but described for example how Chechen
'bandits' fleeing from Grozny shot dozens of their own citizens in Alkhan-Yurt
as punishment for being loyal to Russia. They provoked Russian federal forces
to destroy the village for 'educational purposes'.[76] Other papers remained
'neutral' but quoted official statements denying the crimes extensively and
without critical comment.[77] In such accounts the military campaign was
represented as an orderly and successful process. The *zachistki* of Grozny
(including Staropromyslovsky region) following the bombing was portrayed as a
rational way of 'cleansing' Grozny of more than 1,500 'bandits' so that the
'bearded throat-cutters had no chance of raising their heads and creeping out of
their underground hiding places'.[78] Some of the reports documenting atrocities
and war crimes in Chechnya by the Russian human rights organization
Memorial, and even HRW reports were mentioned in some Russian newspapers.
But they appeared in marginal publications and often long after the events in
question had taken place.[79]

This general pattern of reporting on *zachistki* in Russian newspapers did not
change after the winter of 2000.[80] Several abusive *zachistki* that were well
documented by human rights organizations (such as those in Alkhan-Kala and
Chernorechye 2001) remained entirely beyond the media focus. The *zachistki* of
Sernovodsk and Assinovskaya in early July 2001 were more widely covered, but
did not trigger any significant changes in the patterns of representation.[81] Many
accounts still did not even 'recognize' any guilt of the Russian forces and 'hid'
these events in the discourse on Chechnya as an existential threat to Russia. For
instance, the newspaper *Rossiya* printed an article expressing doubt that any
atrocities had been carried out by Russian soldiers in these villages, noting that
these *zachistki* were a response to a 'terrorist act' and that the federal forces were
not carrying out 'document checks ... for their own pleasure'. Instead, the
second half of the article described the alleged 'ethnic cleansing' of 12,000
Cossacks in these villages in the mid-1990s by the forces of Chechen President
Dudayev.[82]

Russian TV and radio did broadcast several interviews with pro-Moscow
Chechen leaders detailing the abuses and condemning the *zachistki* in
Sernovodsk and Assinovskaya (yet, these accounts appeared in the most
independent' outlets such as Ekho Moskvy and TV6). Several newspapers
covered the *zachistki* and certain more mainstream papers (including *NeGa*)
even criticized the unlawful activities of the Russian forces. In the main,

however, *Izvestiya*'s conclusion that 'cleansing operations in population centres are a natural way for the military authorities in Chechnya to control the territory' represented the dominant position in Russian media on these events, and shows how self-evident the necessity of using emergency measures 'beyond rules that otherwise have to be obeyed' in Chechnya had become over the years.[83] Similarly, the mainstream Russian press presented 'filtration' and 'filtration camps' as a necessity, given the pervasiveness of the terrorist problem – not as something unacceptable.[84]

We can conclude then that potentially 'shocking events' on the battleground in Chechnya did not result in the emergence of a new 'discourse of reconciliation' in the Russian media. There were very few changes to the general pattern of representing the 'Chechen/terrorist' side as different and dangerous, with 'Russia' as the righteous defender. Images and representations that gave the Chechen population an identity as the suffering victims of Russian violence, let alone a human face, did not (re-)appear.

Neither did such changes in the discourse appear in statements by the Russian political elite. Typical examples of statements by Duma representatives triggered by reports about civilian casualties in Chechnya were: 'Of course, it would be desirable if terrorists could be subdued without any damage to the civilian population, which, unfortunately, is not possible given the magnitude of the Chechen problem'[85] or 'the measures taken by the state should be adequate to the extent of terrorism. And ours are adequate'.[86] The prevalence of such statements attests to the continued acceptance in this part of the Russian audience of the official securitization of the Chechen threat far into the war.[87]

Reviewing official representations in connection with potentially 'shocking events' such as the *zachistki* in Alkhan-Yurt, Staropromyslovsky and later in Assinovskaya and Sernovodsk as well as the filtration camp Chernokozovo reveals a pattern of denial combined with justifications for any admitted abuses. Justifications were buttressed with references to the core components of the discourse on Chechnya as an existential terrorist threat, as articulated in the official securitizing narrative from the very beginning.

Allegations and documentation by Malik Saydullayev (head of the pro-Russian Chechen State Council) that Russian forces had committed atrocities in his home village Alkhan-Yurt and that 18 servicemen had been detained in connection with these events were dismissed by the General Procuracy in the Northern Caucasus.[88] The civilian deaths in Alkhan-Yurt were instead blamed on the Chechen fighters.[89] Even the former Chechen mufti Akhmed Kadyrov, who was appointed head of the pro-Russian Chechen administration by President Putin in June 2000, dismissed mass atrocities by Russian forces as 'rumours' or 'disinformation' spread by the fighters in order to harm the authority of his administration. According to Kadyrov 'Single negative facts were presented as a system by local, Russian and international media'.[90] That was also the mantra of Vladimir Kalamanov, Special Representative of the President of the Russian Federation, for ensuring human and civil rights and freedoms in the Chechen Republic, who referred to the breaches of human rights in Chechnya as being 'episodic' and 'exceptions'.[91]

The events at the filtration camp at Chernokozovo could have provided all the necessary ingredients for an official *mea culpa* by the Russian authorities; indeed, a 'total makeover' of the camp was conducted by the Russian authorities before international investigators came to inspect the site. However, official linguistic representations on Chernokozovo *never* linked atrocities and brutalities to Russian servicemen.[92] Even Vladimir Putin's newly-appointed Special Representative for human and civil rights and freedoms in Chechnya Vladimir Kalamanov dismissed the allegations of gross abuses at Chernokozovo: 'it is a glaring lie to portray Chernokozovo as a place where people are shot and tortured almost every day'.[93]

Instead, and over time, official statements on Chernokozovo served to reify representations of the Russian engagement in Chechnya as a humanitarian mission. According to the Prosecutor General of Russia Vladimir Ustinov the 'conditions at the SIZO in Chernokozovo are under the constant surveillance of the Procuracy. All detainees are supplied with bed sheets, three hot meals a day and medical help'. Allegations of atrocities against civilians in Chechnya were 'unfounded' and 'subjective'. Russian military servicemen were 'giving their lives, so that the terrorists would not enter other territories, including the European'. At the same time, the difference and danger of the 'terrorists' or 'bandits' were alluded to, thus legitimizing the practices of Russian servicemen in Chechnya: 'The fighters have rights according to the law and we don't allow encroachments of their rights. But a gracious relation to these bandits is not possible and will not be.'[94]

The widely exposed July 2001 *zachistki* in Assinovskaya and Sernovodsk elicited statements by Akhmad Kadyrov that broke radically with the core identity attached to Russia in the dominate securitizing narrative: 'The counter-terrorist operation is now directed against the peaceful population, not the bandits ... our efforts to help stability and create conditions for the return of refugees have been thwarted by ill-conceived and criminal actions.'[95] Also Russia's top military commander in Chechnya General Vladimir Moltenskoy at first admitted 'large-scale crimes' and 'lawless acts' by Russian forces in Assinovskaya and Sernovodsk.[96] But attaching labels such as 'criminal' to Russian forces found no wider resonance in official statements and was bluntly rejected elsewhere. Russian Interior Minister Boris Gryzlov's initial response to the allegations was that *zachistki* 'should be conducted and they are conducted with respect to the law regulating counter-terrorist operations'.[97] General Moltenskoy retracted his initial statement and said: 'I am unable to speak about crimes. I speak about violations at the level of ordinary soldiers or militiamen; everything was carried out in line with these plans; but some violations were committed.' After a preliminary investigation had been conducted, Moltenskoy said that most residents had provided 'nothing to confirm them (the allegations)'. The heavy-handed tactics used during the *zachistki* had been provoked by the civilians themselves.[98]

Presidential statements followed the same line of reasoning: not total denial of the fact that there had been 'irregularities or abuses', but legitimizing them with reference to their being 'perhaps an inevitable consequence of the battle against terrorism'.[99] In his first formal news conference open to all journalists in July 2001, Putin explained the *zachistki* in Assinovskaya and Sernovodsk a few weeks earlier as follows:

> One of the tactics of the radical fundamentalists that are still trying to operate on Chechen territory is to deliver terrorist attacks against the federal forces, on the one hand; on the other, to attempt to provoke a response attack and put the local population under this attack in order to rouse the local population against the federal authorities. Well, the so-called combing operations which you have mentioned essentially boil down to passport checks and measures to identify the people who are on the federal wanted list. I am not sure that the federal authorities always succeed in not yielding to the provocations staged by fighters. I have been saying this repeatedly and I can repeat it again: all that is being done against the law and against civilians should be exposed and those guilty should be punished.[100]

Despite the reference to the law and to punishment of those guilty, this 2001 statement by the Russian President takes us full circle. It shows how the core components of the official securitizing narrative on Chechnya presented during summer/autumn 1999 were invoked throughout the war and served to legitimize apparently unacceptable practices of war while and after they were undertaken.

Conclusion

This chapter has shown how the dominant discourse on Chechnya as an existential terrorist threat to Russia was enacted in material practices directed both at Chechen fighters and Chechen civilians during the ground offensive. More generally, the securitizing narrative and the very urgent situation that reiterations of this narrative contributed to establish served to legitimize a whole set of emergency measures that broke with both legal and social rules in Russia.

The *zachistki*, with all the violence that accompanied such operations, became a key practice of warfare during the Second Chechen War. It was conducted in a systematic fashion, becoming normal and routine as time passed. Also, the practice of arbitrarily detaining and subjecting people to violence at 'filtration points' acquired a systematic and mass character during the war. Although these practices were not entirely new, they were taken to new heights compared to the First Chechen War; certain aspects of these practices were even institutionalized. The core interest guiding the exploration of these practices in this chapter has been how such systematic yet arbitrary use of violence became legitimate, and how language was important in this process.

Language mattered in at least three different ways – all of them being different expressions of a post-structuralist approach to securitization. First, the broad securitizing narrative offered at the outset of the campaign made practices that would serve to 'cleanse' and 'filtrate' the entire Chechen territory seen logical and necessary. Moreover, the words in the official securitizing narrative that described the 'way out' stipulated the way this should be done. Putin's infamous pledge, 'we will waste them in the can',[101] found material expression in the brutality and violence employed by Russian forces on the ground in Chechnya. Even the targeting of Chechen civilians was to some extent foregrounded in linguistic representations during autumn 1999 which failed to distinguish between 'Chechens' and 'terrorists'. Attaching an extreme level of threat to this entire group of people made possible highly discriminatory practices such as indiscriminate violence against Chechen civilians on the battlefield in Chechnya.

Second, the official securitizing narrative and even very specific aspects of it were echoed in the language of top military personnel. This served not only to reinforce the discourse on Chechnya as an existential terrorist threat to Russia with yet another layer in the Russian public space: it also brought it closer to and passed it onto the soldiers who were to carry out the operations. General Viktor Kazantsev's statement in January 2000, 'We will cleanse Chechnya of any scum', was not a direct instruction on how to conduct *zachistki*, but it reiterated the very broad, vague and de-humanized target and certainly did not caution against arbitrary violence. It rather seemed to indicate that such violence might be necessary.

Moving down to the micro-level, the language used by various kinds of security personnel in executing these practices included core components of the official securitizing narrative. The enactment of violent practices on the ground in Chechnya or at a police station in Moscow was accompanied by linguistic representations that were linked to and referred to the dominant discourse of Chechnya as an existential terrorist threat and Russia as the righteous defender. Even specific de-humanizing nouns from the official securitizing narrative can be found in the language of Russian soldiers committing atrocities in Chechnya.

Third, the thick discourse on Chechnya as an existential terrorist threat, established before the ground offensive started, served to carry and cover potentially 'shocking events' as they took place. There were no words or pictures that could alter the position of 'Russia' as the 'righteous defender' and 'Chechnya' as the 'existential terrorist threat', and thus no grounds for the re-emergence of the 'discourse of reconciliation' in Russian media. Nor were there in the Russian political elite or expert statements on such 'shocking events'. If Russia was identified with any sort of brutality or atrocity, this was represented (with reference to the threat) as a necessity. As for the official discourse, comments on the potentially 'shocking events' in Novye Aldy, Chernokozovo or Assinovskaya took the form of a consistent insistence on sticking to the initial securitizing narrative. Even the cruellest instances of abuse by Russian soldiers during the counter-terrorist campaign in Chechnya were most often rephrased by Russian officials in line with core components of the initial securitizing narrative.

Paradoxically then, the enactment of linguistic representations of Chechnya over time, in material practices served to reinforce the initial categorizations of

Chechnya and Russia even when the potentially 'shocking' results of these practices emerged. This was not only so because of the way these events were handled linguistically. The continuous material practices of 'cleaning', bombing, detaining and sealing off Chechnya and Chechens from Russia, interpreted through the lens of the securitizing narrative, served to constitute and maintain Chechnya and the Chechens as different and dangerous, with Russia as the righteous defender.

The underlying claim throughout this chapter and the two previous ones has been that such constant reiteration of representations of Chechnya and Chechens as different and dangerous hardened the discourse to such an extent that warfare practices that broke sharply with legal norms of armed combat (and also with what one would think are social norms constraining what a person may do to fellow human beings) appeared 'normal' and appropriate. Admittedly, the broader Russian audience could not 'see' all the violence practised in Chechnya and against Chechens – but it still seems puzzling that the information they did get failed to spur broader reactions against these atrocities. Public opinion polls in 2000 showed that 87 per cent of the Russians surveyed were convinced that 'only the Chechens themselves were to blame for the military conflict'. Only 22 per cent believed information in the Russian media about the brutality and impunity practised by Russian forces in Chechnya.[102]

The reception of the Budanov case in Russian society is a case in point. Yury Budanov was a decorated tank commander responsible for abducting, raping and killing a young Chechen girl. He was one of few servicemen actually accused of his crimes during the Second Chechen War. Chief of the Russian General Staff General Anatoly Kvashnin even went on Russian television to denounce the killing as 'barbaric and disgraceful'. And yet, Budanov became a hero among the Russian public. During the trial, crowds gathered outside the courthouse to demand his release. In a public opinion poll conducted by the Public Opinion Foundation at the time of his trial in April 2001, 50 per cent of the respondents wanted the trial stopped and Budanov released.[103]

How was this possible? My argument has been that language matters for understanding how gross human rights violations during war can become acceptable.

To return to a core theory assumption presented at the outset, the legitimacy of a policy or practice indicated as a 'way out' in the securitizing narrative rests on its congruence with the level of threat implied in the representation of the threat. There was congruence between the level of threat implied in linguistic representations of Chechnya and the violence undertaken against Chechnya/ Chechens during the Second Chechen War. That is how the war became acceptable – both to those conducting it and among the Russian public.

Notes

1 Rachel Denber, cited in *RFE/RL Newsline*, 6 August 2001.
2 The torture and abuse documented in the various reports noted below are seriou‹

violations of Russia's obligations under the Geneva Conventions of 1949 and Protocol II to the Convention, which elaborates the rules for internal armed conflict, and under the instruments of international human rights law to which Russia is also party.

3 'Yest' argumenty ubeditel'ney puli', *Trud*, 5 October 1999.
4 'Na Chechenskom fronte bez peremen', *Vremya MN*, 23 November 1999.
5 'Taynyye i yavnyye manevry Moskvy', *NeGa*, 5 October 1999.
6 Referred to in document by Memorial available at www.memo.ru/2008/09/04/0409081eng/part5.htm and accessed 17 March 2016.

7 Such as ... 2) to check the identity documents of private persons and officials and, where they have no identity documents, to detain them for identification; 3) to detain persons who have committed or are committing offences or other acts in defiance of the lawful demands of persons engaged in an counterterrorist operation, ... ; 4) to enter private residential or other premises ... and means of transport while suppressing a terrorist act or pursuing persons suspected of committing such an act, when a delay may jeopardise human life or health; 5) to search persons, their belongings and vehicles entering or exiting the zone of an counterterrorist operation, including with the use of technical means; ...

(FZ 'On the Suppression of Terrorism' (1998), Article 13)

8 Only limited measures were eventually taken to encourage or discipline the civilian and military procuracy to end the climate of impunity in Chechnya: Order No. 46 was issued by the Prosecutor General of Russia and then followed up by Order No. 80 issued by General Moltenskoy in March 2002 ('Decree/Order No. 80 of the Command of the United Group of Forces in the Northern Caucasus Region of the Russian Federation, on Measures to Enhance Efforts by Local Governmental Authorities and Law Enforcement Agencies of the Russian Federation in the Fight Against Unlawful Actions and Accountability for Officials for Violations of Law and Law and Order in the Conduct of Special Operations and Targeted Operations in Settlements in the Chechen Republic.' Issued March 27, 2002, Khankala).
9 See for example Human Rights Watch. 2002. *Last Seen ... : Continued 'disappearances' in Chechnya*: 6. Available at www.hrw.org/reports/2002/04/15/last-seen-0 and accessed 17 March 2016.
10 Human Rights Watch. 2000d. *No Happiness Remains. Civilian killings, looting, and rape in Alkhan-Yurt, Chechnya*. Moscow: Memorial. Available at www.hrw.org/reports/2000/russia_chechnya2/index.htm and accessed 17 March 2016.
11 Human Rights Watch. 2000e. *Civilian Killings in Staropromyslovsky District of Grozny*. Available at www.hrw.org/reports/2000/russia_chechnya/ and accessed 17 March 2016.
12 Memorial and Demos. 2007. *Counterterrorism Operation by the Russian Federation in the Northern Caucasus throughout 1999–2006*. Available at: www.memo.ru/hr/hotpoints/N-Caucas/dkeng.htm. Accessed 31 July 2013.
13 For a 'reconstruction' of the *zachistka* of Novye Aldy see E. Gilligan. 2010. *Terror in Chechnya. Russia and the Tragedy of Civilians in War*. Princeton, NJ: Princeton University Press, 54–58. See also Amnesty on mass killings in these three *zachistki* (Amnesty International. 2002. *Denial of Justice*. London: Amnesty International Publications. Available at: www.amnesty.org/en/library/info/EUR46/027/2002).
14 Gilligan 2010, 63–64.
15 Memorial. 2001. *The 'Cleansing' Pperations in Sernovodsk and Assinovskaya were Punishment Operations*. Available at www.memo.ru/eng/memhrc/texts/sern&assin.shtml and accessed 17 March 2016. Memorial. 2002b. *Swept under: Torture, Forced Disappearance, and Extrajudicial Killings during Sweep Operations in Chechnya*. Available at www.hrw.org/reports/2002/russchech/chech0202.pdf and accessed 17 March 2016.

16 See Memorial report on the 25 July 'Cleansing operation in the village of Tsotsin-Yurt' available at www.memo.ru/eng/memhrc/texts/cocinurt.shtml and accessed 17 March 2016. See also Memorial (2002a. *Myths and Truth about Tsotsin-Yurt December 30, 2001–January 3, 2002.* Available at www.memo.ru/eng/memhrc/texts/mythtruth.shtml.) and Amnesty International (2002, 56–57).

17 Memorial. 2008a. *Special Operations [zachistka].* Available at www.memo.ru/2008/09/04/0409081eng/part5.htm and accessed 17 March 2016.

18 Gilligan 2010, 63–64.

19 Memorial 2002b and Human Rights Watch (2002. *Last seen …: Continued 'disappearances' in Chechnya.* Available at: www.hrw.org/reports/2002/04/15/last-seen-0).

20 Musayev and Others v. Russia, (57941/00, 58699/00, and 60403/00) Judgment, 26 July 2007, Estamirov and Others v. Russia, (60272/00) Judgement, 12 October 2006, (these two cases are about extrajudicial execution during the 'mopping-up' operation in Novye Aldy in February 2000; Tangiyeva v. Russia (Application no. 57935/00) Judgment, 29 November 2007; Medov v. Russia, (1573/02) Judgment, 8 November 2007; Chitayev and Chitayev v. Russia, (59334/00) Judgment, 18 January 2007; Ayubov v. Russia, (7654/02) Judgement, 12 February 2009; Musayeva v. Russia, (12703/02) Judgement, 3 July 2008; Amuyeva and Others v. Russia, (17321/06) Judgement, 25 November 2010 (a special operation aiming to identify member of illegal armed groups in the village Gekhi-Chu); Goncharuk v. Russia, (58643/00) Judgement, 4 October 2007 (extrajudicial execution during an attack on the Staropromyslovsky district in January 2000). (Available at www.srji.org/en/legal/cases To find the summary of the case plus link to the full text put the name of the applicant into the 'text search').

21 J. Russell. 2007. *Chechnya – Russia's 'War on Terror'.* London: Routledge, 84.

22 See for example the testimonies cited in Human Rights Watch (2000b. *Welcome to Hell.* Available at: www.refworld.org/docid/3ae6a8750.html, 13–15).

23 Human Rights Watch. 2000c. *Hundreds of Chechens detained in 'filtration camps'. Detainees face torture, extortion, rape.* Available at www.hrw.org/news/2000/02/17/hundreds-chechens-detained-filtration-camps and accessed 17 March 2016.

24 Memorial and Demos 2007, 24; see also Memorial 2002b.

25 Memorial and Demos 2007, 25.

26 Memorial 2002a. *Myths and Truth about Tsotsin-Yurt December 30, 2001–January 3, 2002.* Available at www.memo.ru/eng/memhrc/texts/mythtruth.shtml and accessed 17 March 2016.]; Memorial 2002b; Human Rights Watch 2000b; Amnesty International 2002.

27 Memorial 2002a; Memorial 2002b.

28 Six people disappeared after being detained at the 'temporary filtration point' in Tsotsin-Yurt (Memorial 2002a). According to Memorial more than 4000 Chechens 'disappeared' from the beginning of the war in 1999 until 2004 ('The Chechnya vanishing point: Fate of thousands unknown', *AFP*, September 28, 2004).

29 Human Rights Watch. 2000c. *Hundreds of Chechens Detained in 'Filtration Camps'. Detainees face torture, extortion, rape.* Available at www.hrw.org/news/2000/02/17/hundreds-chechens-detained-filtration-camps and accessed 17 March 2016; Amnesty International UK. 2000a. *Rape and Torture of Children in Chernokozovo 'Filtration Camp'.* Available at www.amnesty.org.uk/news_details.asp?NewsID=12978 and accessed 17 March 2016.

30 Amnesty International. 2000b. *Real Scale of Atrocities in Chechnya: New evidence of cover-up.* Available at www.amnesty.org/fr/library/asset/EUR46/020/2000/en/3a33f6e5–3e4d-407f-b3c6–9f1b60ce584a/eur460202000en.pdf and accessed 17 March 2016.

31 Memorial. 2008b. *Filtration system.* Available at www.memo.ru/2008/09/04/0409081eng/part61.htm and accessed 17 March 2016. Human Rights Watch. 2000b *Welcome to Hell*: 34–36. Available at www.refworld.org/docid/3ae6a8750.html and accessed 17 March 2016.

32 Memorial 2008b.
33 Amnesty International 2000b. For testimonies on torture and use of the 'live gautlet' at detention centres, see Human Rights Watch 2000b.
34 Under Russian law, persons suspected of having committed crimes of a terrorist nature or of participation in illegal armed formations are to be transferred to organs of the Prosecutor's Office or the FSB, not delivered to a place where a military unit is deployed.
35 Memorial 2008b.
36 The bodies reportedly bore signs of torture and mutilation (Amnesty International 2002, 63).
37 Human Rights Watch 2000b, 29–33.
38 Memorial and Demos 2007, 25.
39 Memorial 2008b.
40 'Na Lubyanke znayut, kto sovershil terakty', *NeGa*, 25 September 1999.
41 Boris Yeltsin cited in 'Boris Eltsin upovayet na silovikov', *NeGa*, 17 August 1999 and Vladimir Putin cited in 'Duma dala Vladimiru Putinu neobkhodimoye', *NeGa*, 17 August 1999.
42 Vladimir Putin cited in 'Nado zadushit' gadinu na kornyu', *RoG*, 17 September 1999, or his lengthy interview with editors of regional newspapers in 'Vladimir Putin: "Chechnya zanimaet tol'ko 45% vremeni v rabote pravitel'stva', *Chas Pik*, 20 September 1999, or Defence Minister Igor Sergeyev quoted in 'V Kreml' cherez Chechnyu?', *Segodnya*, 28 September 1999.
43 Cited in 'Voyna', *Russkiy Dom*, 3 January 2000.
44 Cited in 'V Chechne nachat vtoroy etap voyskovoy operatsii', *NeGa*, 19 October 1999.
45 When casualty figures were given by the headquarters of the federal troops in the North Caucasus, those killed were routinely referred to as 'terrorists' or 'extremists' (See for example *Itar Tass*, Grozny, 10 April 2001).
46 General Troshev quoted in 'S Arguna nachalas voyna', *NeGa*, 8 December 1999. See also 'Samoye trudnoye – vperedi', *NeGa*, 27 October 1999.
47 'Blokanyy Grozny', *NeGa*, 29 October 1999. According to General Vladimir Shamanov, officers repeatedly reminded the soldiers that the terrorists employed methods that had nothing in common with the art of war. (Interview with General Vladimir Shamanov in 'Rossiya ne dolzhna opravdyvat'sya', *NeGa*, 4 November 1999).
48 'Protivostoyaniye v Chechne blizitsya k kul'minatsii', *NeGa*, 6 November 1999.
49 'A former rebel tells of his war in Chechnya', *AFP*, 2 October 2004.
50 Quoted from 'Chelovek iz Drugogo ushchel'ya: Beseda v bronetransportere s nachal'nikom razvedki po doroge na Duba-Yurt', *Izvestiya*, 28 March 2003, referred in Gilligan (2010, 59).
51 *RFE/RL Newsline*, 19 August 1999.
52 Referred from the killings of civilians in Novye Aldy i (February 5; a day of slaughter in Novye Aldy posted at www.hrw.org/reports/2000/russia-chechnya3/Chech006–05.htm and accessed 17 March 2016.)
53 Human Rights Watch 2000b: 11.
54 'Za Terekom tozhe nasha zemlya', *RoGa*, 15 October 1999.
55 Numerous interviews with former detainees in connection with *zachistki* in Alkhan-Kala (19–25 June 2001) Sernovedsk (2–3 July 200) and Assinovskaya (3–4 July 2001), referred to in Memorial (2002b).
56 Human Rights Watch 2000b, 20.
57 Testimony by 'Akhmed Isaev' (Human Rights Watch 2000b, 21).
58 Testimony of 'Sultan Denoev' (Human Rights Watch 2000b, 34).
59 Interview with Zura Davletukaeva, Nazran, Ingushetiya, 12 July 2001 cited in Memorial (2002b, 20).

60　Putin several times referred to them as 'animals' ('Terrorists are people, not animals', *Moscow Times*, 21 March 2000) or 'rabid animals' ('Moscow awash in explosion theories', *Moscow Times*, 14 September 1999); Yeltsin called them 'wild beasts' ('Moscow awash in explosion theories', *Moscow Times*, 14 September 1999).

61　Testimony by 'Saipudin Saadulayev' (Human Rights Watch 2000b, 32). Gilligan (2010, 73) also notes how words such as 'apes', 'cattle', 'wolves' and 'dogs' were used to refer to Chechens during torture.

62　Human Rights Watch. 2000d *No Happiness Remains. Civilian killings, looting, and rape in Alkhan-Yurt, Chechnya*. Moscow: Memorial. Available at: www.hrw.org/reports/2000/russia_chechnya2/index.htm.

63　Sniper Vyacheslav Kravets, interviewed in 'Moskovskoye myaso', *Sovershenno Sekretno*, 11 July 2000.

64　'Nam pokazali, kak nas nenavidyat', *Moskovskye Novosti*, no. 31 (31 July–6 August).

65　Memorial 2002a. There are also other examples of torture and ill-treatment justified with reference to the victim being 'Chechen'. For example, testimony of 'Badrudy Kantaev': 'They beat me terribly ... They would punch you and say, 'You damn Chechen, why aren't you falling over' (cited in Human Rights Watch 2000b, 31).

66　Testimony of 'Sultan Deniev' (Human Rights Watch 2000b, 13).

67　Human Rights Watch. 2000d.

68　Testimony of 'Elisa Ebieva', p. 12; testimony of 'Yakub Tasuev', p. 19; testimony of 'Ali Baigiraev', p. 22 in Human Rights Watch (2000b).

69　Testimony of 'Abdul Jambekov', p. 18; testimony of Movsar Larsanov, p. 21 in Human Rights Watch (2000b).

70　Eyewitness to *zachistka* in Novye Aldy Larisa Labazanova, referred in 'Chechen Massacre Survivors See Justice' by Asya Umarova, *Caucasus CRS* Issue 405, 18 August 2007, available at http://iwpr.net/report-news/chechen-massacre-survivors-see-justice and accessed 17 March 2016.

71　The chronicles of events printed in *NeGa* in October listed the number of bomb attacks, cleansing operations, ('zachistki'), and the number of killed fighters. Only one incident of civilian casualties was noted and this incident was attributed to the fighters of the Chechen warlord, Musa Mezhidov. Only one war-crime was noted, which was attributed to Chechen fighters ('Severnyy Kavkaz: khronika konflikta', *NeGa*, 16 October 1999 and 'Chechnya: Khronika konflikta', *NeGa*, 30 October 1999). Very similar chronicles were posted in *RoGa* during October.

72　The reporting from 15–19 October can serve as an example: 15 October 'Yesterday the Federal Forces met active resistance in the region of Goragorsk, and they began annihilating the Chechen fighters' ('Miting v Groznom', *NeGa*, 15 October 1999), 16 October (first page):

　　A third of Chechnya has been liberated ... first phase of the counterterrorist operation had cost 112 lives among Russian servicemen ... at least 1500 bandits had been destroyed.... The Federals yesterday started to 'cleanse' Goragorsk, where a large group of fighters were concentrated.

　　　　　　　　　　　　　　　　　('Osvobozhdena tret' Chechni', *NeGa*, 16 October 1999)

　　'In Chechnya the second part of the military operation is underway ... Russian forces are advancing to destroy the terrorists on the entire Chechen territory' ('V Chechne nachat vtoroy etap voyskovoy operatsii', *NeGa*, 19 October 1999).

73　See for example 'Obstanovka v Severo-Kavkazskom regione na 19 Oktyabrya' *Kasnaya Zvezda*, 20 October 1999; 'Chechnya: khronika konflikta', *Nezavisimoye Voyennoye Obozreniye*, 22 October 1999, 5 November 1999 and 14 January 2000 'Na Chechenskom fronte bez peremen', *Vremya MN*, 23 November 1999; 'Severnyy Kavkaz: Khronika konflikta', *NeGa*, 16 October 1999 or 13 November 1999.

74 'Chechnya: Khabarovskiy OMON kak mirotvorets iz OON', *Tikhookeanskaya Zvezda*, 12 January 2000. Another interview with returning OMON soldier that carries similar representations is 'OMON predotvratili perevorot', S*hchit i Mech*, 3 February 2000.
75 'Sam ne videl, no govoryat', *Trud*, 15 December 1999, 'Gornyye boi v Chechne', *NeGa*, 25 December 1999 and 'Kapkan zakhlopnulsya', *Gudok*, 28 December 1999.
76 'Pobedy vysokaya tsena', *Ekspert*, 28 February 2000.
77 'Grozny vzyal Rossiyskiy zek', *Kommersant*, 25 December 1999.
78 'Kapkan zakhlopnulsya', *Gudok*, 28 December 1999.
79 'Eto strashnoe slovo 'zachistka''', *Russkaya Mysl'*, 14 September 2000; 'Sledstviye ne zakoncheno', *Novyye Izvestiya*, 26 December 2000 (Still memorial); 'Prikaz dlya OON ili dlya Rossii?', *Novye Izvestiya*, 2 April 2002; 'Tak nazyvayemaya zachistka', *Vremya Novostey*, 5 March 2000; 'Kontrol'naya zachistka', *Vremya Novostey*, 12 May 2000; and 'Polozheniye v Chechne ukhudshilos', *Novyye Izvestiya*, 21 February 2001.
80 'Boyeviki idut va-bank', *Gazeta.ru*, 17 Juli 2000; 'Khattab kupil neskol'ko legko-motornykh samoletov, soobshchayut voyennyye istochniki', *DeadLine.ru*, 11 September 2000.
81 See for example Lidiya Grafova's account of a fact-finding mission to the villages in 'Nam pokazali, kak nas nenavidyat', *Moskovskiye Novosti*, no. 31 (31 July – 6 August 2001).
82 'Russkiye Kazaki obideli Chechenskikh stanichnikov', *Rossiya*, 10 July 2001.
83 Referred to in 'Russian media mull Chechnya abuses', *BBC News*, 11 July 2001.
84 'Federaly vyshli k Tereku', *Vechernyaya Moskva*, 6 October 1999; 'Pobedy vys-okaya tsena', *Ekspert*, 28 February 2000.
85 Konstantin Kosachev of the Otechestvo Party, quoted in 'Vlast' ne vyderzhala ispy-taniya oppozitsiyey', *NeGa*, 17 February 2000. Translated into English by S. Mäkinen, available at http://tampub.uta.fi/handle/10024/67815).
86 Boris Gryzlov of the Edinstvo party quoted in *Vek*, 10 February 2000. Translated into English by S. Mäkinen 2008, 204.
87 There were some critical voices, such as Yabloko Party leader Grigory Yavlinsky and former CIS secretary Boris Berezovsky, but these were few ('Mirotvorcheskiye voyska', *NeGa*, 1 December 1999).
88 The investigation into the *zachistka* in Novye Aldy finally concluded that no crimes had been committed ('Pomoshch' PACE v rassledovanii prestupleniy v Chechne ne nyzhna', *Strana.ru*, 22 March 2001).
89 'Grozny vzyal Rossiyskiy zek', *Kommersant*, 25 December 1999. On a similar episode in June 2000 see 'Vse, kak v Afgane, tol'ko geroyev bol'she', *Vechernyaya Moskva*, 8 June 2000.
90 Interview with Kadyrov in 'Chechentsev slovami uzhe nikto ne kupit', *Novyye Izvestiya*, 17 October 2000.
91 'Polozheniye v Chechne ukhudshilos', *Novyye Izvestiya*, 21 February 2001.
92 'Russia rattled by torture claims at Chechen camps', *Independent*, 18 February 2000; 'Russian Justice Ministry Denies Atrocity Reports', World News Connection, *Itar-Tass*, 26 February 2000; 'Angry Russia defends its rights record before Washing-ton', *Agence France-Presse*, 1 March 2000.
93 'Kalamanov says no filtration camps in Chechnya', *Itar-Tass*, 1 March 2000. See also 'Human rights commissioner denies Chechnya torture reports', *Interfax News Agency*, 29 March 2000.
94 Quotes by Vladimir Ustinov in response to allegations by PACE representatives that atrocities against civilians in Chechnya were not being properly investigated, referred in 'Pomoshch' PACE v rassledovanii prestuplenii v Chechne ne nuzhna', *Strana.ru*, 22 March 2001.
95 'Moscow-appointed administrator of Chechnya accuses Russian forces of crimes', *Associated Press*, 10 July 2001.

96 Memorial 2002b: 43. Indeed, as a sign of recognition, a special commission was appointed to investigate the activities of Russian troops in the villages of Assi-novskaya and Sernovodsk ('How to prevent the issue of Chechnya from coming up at the G-8 summit', *Rossiya*, 19 July 2001).

97 'Interior Minister says Chechnya cleaun-ps are lawful', *Interfax News Agency*, 6 July 2001.

98 Memorial 2002b: 43.

99 Vladimir Putin, referred in *Corriere della Sera* 'Putin describes Chechen abuses as "inevitable"', *Agence France Presse*, 16 July 2001.

100 BBC Monitoring 'Russian president gives extensive news conference in the Kremlin'. Source: Russia TV, Moscow, in Russian 1300 gmt 18 Jul 01 as carried on *Johnson's Russia List*, 19 July 2001.

101 'Voyna bez vykhodnykh', *NeGa*, 25 September 1999.

102 Poll, referred to in 'Eshche odna voyennaya zima v Chechne', *Russkaya Mysl'*, 7 December 2000.

103 Poll referred to in 'Russian colonel hailed as hero for killing of Chechen woman', *Sunday Telegraph*, 15 April 2001.

13 Conclusions and perspectives

With this book I have sought to understand how war becomes acceptable. The empirical puzzle at the centre of this endeavour has been how a second post-Cold War military campaign by Russia against Chechnya became a legitimate undertaking. Not only had the first campaign against this tiny Russian republic turned out to be a totally unacceptable enterprise, a second campaign was unthinkable for most Russians only months before it was launched in autumn 1999. Yet, the Second Chechen War, a war just as violent as the First Chechen War, was acceptable, even required, in the eyes of the Russian public at the outset and as the war dragged on.

The main part of this concluding chapter will recap the findings on how the social process that made the war acceptable unfolded and provide some broader perspectives on Russia and the consequences of this so-called counter-terrorist campaign. However, given the ambition of the book to also say something about war and legitimation in general, I wish to start with presenting some claims about how securitization works before and during war.

First, rhetorical preparations before war matter. Many wars are launched on a weak rhetorical foundation. They might still be launched and fought, as was the First Chechen War, but support for such a war is weak in the first place and easily dwindles. By contrast, other wars – such as the Second Chechen War – become acceptable. They are launched accompanied by a thick securitizing narrative that is *consistent*, in the sense that the representation of the enemy can be placed at the top of a scale in terms of danger and matches the policy of war that is indicated.

Second, discursive context matters. Official securitizing attempts before and in war can acquire legitimacy if they draw on ingrained and established representations of threat in the national discursive terrain and/or on a threat representation that is dominant in the international discursive terrain at the time. It is easier to fight an acceptable war against someone if this 'someone' has historically been constructed as different and dangerous and/or if the classification in which this someone is placed is particularly salient at the time. It might not matter if the war is fought close to home or far away, but it does matter how this someone has been represented over time and how predominantly this representation figures in the national (or international) discourses of threat as such. There are such things as habitual enemies: choosing them as an object of war contributes to making the war acceptable.

A third and perhaps controversial claim about securitization and war is that when war becomes acceptable this is thanks to the discursive efforts of many. Both what Buzan and Wæver refer to as the 'securitizing actor' *and* 'the audience' contribute. Guzzini has claimed that 'many analysts (and innumerable student papers) fall prey to reducing securitization to studies in which they simply expose the intentional war-mongering of some political actors'.[1] This study indicates that, by emphasizing securitization as an intersubjective process of legitimation, the spotlight is broadened beyond the war-mongering leadership and can shed light on how the political opposition, experts, generals, police and especially the media not only *accept* but *contribute* to the construction of the object as an existential threat and to making the war a legitimate undertaking. If we want to assign responsibility for war and violence, it is important to recognize the role played by 'actors' other than the political leadership. After all, these actors always have some possibility of voicing representations that counter those depicting the object as an existential threat.

Fourth, this study indicates that the type of classification/representation agreed upon during the process of securitization effects how that war can be waged. While it is impossible to rank different wars according to degree of 'cruelty' along an objective standard, some wars are clearly more violent than others in terms of how massive and indiscriminate the violence is, and how long it can be carried out and still be acceptable. Securitizing narratives in war that cast the enemy as extremely dangerous and different make massive and indiscriminate violence possible and acceptable. Further, such acceptance must be nurtured as the war rolls on. Securitization is never a stable social arrangement; neither are acceptable wars. Discourses that negate the image of the enemy as different and dangerous, and represent the victims of war as fellow human beings can emerge to challenge this representation. In particular, a war that entails heavy human costs must be constantly legitimized through representations of the enemy as an existential threat. Again, continued acceptance is not necessarily the work of the political leadership alone: it is better seen as a collective endeavour where the entire potential audience plays a role. Scaling down threat representations can always be undertaken, and it may start among the putative 'audience'.

Finally, securitization for and in war creates conditions not only for acceptable war, but also for re-drawing the identity of the referent object. The urgent focus and discursive detailing of the threat which a securitizing attempt in war can elicit will also produce a new articulation of the Self that is said to be threatened. It might be argued that this re-articulated Self in time of war is *negatively* constituted, that it is more through what it is *not* than through what it *is* that the Self becomes re-defined and united. Nevertheless, no social group wages an acceptable war and remains the same. There will always be some benefit in terms of social cohesion. But, as I will turn to below in my discussion of the empirical case studied throughout this book, such cohesion may be precarious and come at a price.

A stand against inevitability

In the introduction I promised that this would to be a *critical* but also a *constructive* endeavour in the sense that it would not only reveal how war becomes acceptable but also indicate how war can be replaced by peaceful interaction. Starting the empirical enquiry of Russo-Chechen relations in Chapter 4 with a detour back to the interwar period (1996–1999) was a move in this direction. Not because these were years without problems and violence in Chechnya (this was a period fraught with internal strife and violence), but because in official Russian statements Chechnya was given an identity as a partner and a potential friend. This broke with the historical pattern of representing Chechnya as a radical and dangerous Other in some form or another. This 'positive' articulation does not appear to have become an ingrained or widespread understanding of Chechnya in Russia. But at least this detour has shown that such a change and the policies of cooperation it logically entails are indeed possible.

Some would hold that the relationship between Ramzan Kadyrov's Chechnya and today's Russia is one of friendship and peace. I would argue that this is a mutual friendship that, if it exists at all, is confined to the uppermost level of leadership, and that the relation still hinges on violence. The fact that there are still thousands of federal armed personnel as well as 'Kadyrovtsy' inside and immediately outside the Chechen border testifies to this.[2] For a real change in the relationship to come about, a new articulation of Chechnya and Chechens in Russia is needed, as well as a new articulation of Russia and Russians in Chechnya. The interwar period is a reminder that such articulations are possible.

Let us now return to the critical ambition of this endeavour. According to the basic stand of this book against inevitability, the Second Chechen War did *not have to become* an acceptable war. Clearly, there were forerunners to the discourse on Chechnya as an existential terrorist threat – but it was the formidable accumulation of official statements attaching such an identity to Chechnya during summer and autumn 1999 that brought new urgency to the debate, as documented in Chapter 5. Not only was there a sudden spate of official statements on 'terrorism' and 'Chechnya', but security issues rapidly came to dominate the national agenda as such. The focus on Russia as threatened by economic crises (1998 onwards) or by a weak Yeltsin regime was now, with Putin as Prime Minister, replaced by a focus on an internal/external *violent* terrorist threat in Chechnya.

Thus, it is not without relevance to see just what this securitizing move looked like. The 1999 official securitizing narrative of the terrorist threat was indeed powerful and frightening. The threat was presented as inhuman and capable of gruesome deeds, and at the same time as 'professional', 'well-trained' and with 'far-reaching plans'. With Chechnya as the epicentre of this violent threat, references to links with 'enemy circles in Muslim countries', 'the directors of the terrorist war' and 'Osama bin Laden' heightened the power and omnipresence of the threat even further. Taken together this was a threat construction that could be placed at the top end of the scale in terms of danger and difference. Combined with descriptions of the situation as a *war* declared upon the 'entire

Russian statehood', these official statements conveyed a sense of urgency: 'Russia' was at the point of no return. Now the only 'way out' was violent and uncompromising emergency action that would ensure the 'total annihilation' or 'destruction' of the threat. Not only was there a good fit between the level of threat implied in the description of the threat and the level of violence prescribed in this narrative, it was consistently and persistently repeated in official statements over time.

Although this book has set out to broaden the spotlight *beyond* the warmongering of a political leadership to understand how war becomes acceptable, there can be no doubt that the discursive efforts of the new Russian leadership from August 1999 onward represent a crucial piece in the puzzle. 'Chechnya' was singled out and detailed as an existential terrorist threat. Although there was no declaration of war, the official narrative clearly issued a call for a massive violent undertaking against this republic. A particularly acute problem for the Chechen president and the moderate wing of the Chechen government as well as for the Chechen civilian population was that Russian official statements did *not* clearly distinguish between them and the 'terrorist threat'. They were all subsumed under the terrorist threat. This marked the beginning of a discursive process that would render these groups 'dangerous' and without a human face, their physical lives precarious and dispensable.

Russian identity

The new official statements on Chechnya as an existential terrorist threat served to elevate 'Russia' to a strong and united position. In the larger context of Russian identity formation, this was a significant move. Revisiting Neumann's statement that the larger the group is, 'the more their cohesion depends on some kind of glue, some markers of commonness, some integration',[3] we may note that Russia is a *big* and diverse state, in terms of space as well as population. After the dissolution of the Soviet Union, social cohesion was particularly weak. The Yeltsin regime did not succeed in articulating a new Russian identity that could encompass the many different peoples living on the territory of the Russian Federation. Different nationalisms were flourishing, and central power was weak. This study indicates that the securitization of the Chechen threat from 1999 onward became a vehicle for re-securing the borders of Russian identity in this situation. Indeed, in line with Connolly's[4] theoretical propositions, it might be that the particular challenges facing such a vast and diverse country as Russia in seeking to articulate a positive common Self makes it more prone to focus on the Other, and that the most efficient form of 'Othering' becomes one of radical Otherness. The recurrent invoking of a 'siege-mentality' by the Russian leadership with reference to a Western/U.S. threat[5] and the recent revival of 'terrorist talk' in official statements noted in the introduction to this book would appear to support this interpretation.

Thus, I see securitization as a particularly relevant mechanism for social cohesion in Russia. This book has investigated the comeback of this mechanism in Russian politics. While representations of 'Russia' as strong, righteous and victorious are frequent in the historical discursive terrain explored in Chapter 6, especially in the texts of communist and nationalist opposition in the 1990s, this was not the case in the *official* Russian discourse of the time. As our examination of official interwar representations has shown, 'Russia' was instead depicted as guilty, inadequate and even weak in this period. Moreover, media representations during the First Chechen War had already detailed such a representation.

By contrast, the version of 'Russia' elicited by the 1999 official securitization of Chechnya as a terrorist threat was one of defensive innocence, physical strength, unity and ability to install order. Representations voiced by the political elite, experts and journalists added to these official indications of what Russia is and should be, confirming and expanding on this position, rather than negating it.

With the benefit of hindsight we can see that the re-articulation of Russian identity in the face of the existential terrorist threat in 1999 and 2000 was merely the small beginning of an official re-articulation that was to become increasingly distinct, and has now been supplemented with more 'positive' markers of the Russian Self.[6] Nevertheless, the securitizing narrative accompanying the Second Chechen War and the resonance and amplification that this narrative found in the wider public debate constituted a crucial starting point of this difficult process.

Putin's rise to power

The Second Chechen War not only served as a vehicle for the return and strengthening of a core position on Russian identity – it also produced a surprising re-union of the fragmented Russian political elite under the auspices of the incumbent regime and became a launching pad to power for Vladimir Putin.[7] While Yeltsin had articulated a Russian identity which contradicted that of core constituencies in Russia, we have seen how Putin's securitizing narrative spoke directly to key ideas in the language of the communist and nationalist opposition that always gave Yeltsin trouble. Under the banner of the terrorist/Chechen threat to Russia, diverse groups in the political elite were brought together and linked to the Russian leadership. In line with the new focus on an external/ internal threat to Russia and the call for 'unity', such unification of the Russian polity now seemed reasonable and urgent.

The situation enabled renewed concord among the predecessors to Putin's new position. A meeting could be summoned in the Russian White House to discuss the critical situation in Chechnya which gathered all four former Prime Ministers and the heads of all Duma factions. The most striking feature of this new 'Club of Prime Ministers' was their agreement on the Chechen issue.[8] Russia's situation as a state at war with terrorism and the lifting of this issue to the top of the political agenda also made direct contact with and submission of federal subjects to the Federal Centre logical. Putin indicated as early as September 1999, with reference to the threat facing Russia, that regional security structures – not on paper, but in practice – should be fully and unconditionally subordinated to Federal security structures.[9] A parallel to this came several years later, with President Putin's proposal following the 2004 Beslan hostage crisis to replace direct elections by presidential appointment of regional leaders.

The dramatic rise of the (presidential) Unity Party and the shrinking of the key contender, the (governors') Fatherland-All Russia Party during autumn 1999, must also be re-viewed through the lens of securitization of the Chechen threat. By the end of September, most of the governors had joined the Unity Party, even though it had been considered a liability to be associated with 'the party of power'.[10] The claim is not that the rise of Unity as the dominant party in Russia can be explained by the securitizing discourse: its establishment may to a large degree have come about by pressure, manipulation and command. Nevertheless, the strengthening of a discourse on Russia being 'threatened' and 'at war' and the agreement on how this necessitated a new unity on the political level facilitated the rise of this party in Russia in autumn 1999, making it both logical and legitimate.

What I suggest then is that the 'co-option' of the Duma and the reining in of the Federal Council and the regions that became codified and reinforced during Putin's first presidential period was initially not only and perhaps not primarily a result of pressure and force. The dominance of the securitizing narrative in Russia rendered the incorporation of former independent bastions of power into the 'power vertical' logical and appropriate. The unification of the Russian elite was driven and legitimized by agreement on the gravity of the threat and on Russian *unity* as a required measure for dealing with it. As shown in Chapter 6, this agreement was reached as much by the Russian leadership accommodating to the positions amongst the political opposition (in the interwar period) as by the opposition moving toward the positions held by the leadership.

Beyond the unification of powers in Russia, the new and dominant discourse of danger brought into being a situation that called for urgent action, making certain 'actors' particularly relevant and authoritative to take the lead in the situation facing Russia. The empowerment of the Ministries and Agencies that administer violence through the Second Chechen War has been highlighted several times in this book (Chapters 4 and 10). We can note similar logic at work for Putin. The discursive elevation of the security challenges as early as in connection with the Shpigun abduction and the incursion into Dagestan made Yeltsin's choice of FSB chief Vladimir Putin as Prime Minister logical in the first place (see Chapter 4). Once in position, the continuation and exacerbation of the security situation seemed to call for a strong leader to stand at the helm of a united Russia at war.

The papers noted that Putin appeared in public as 'a Commander in Chief' during his first months as Prime Minister: a commander-in-chief fighting 'real and potential terrorists' at a time when the population was living in fear of new terrorist attacks.[11] And indeed, from the media coverage during these months we may conclude that it projected Putin as the incarnation of the new 'Russia' strong, determined and capable of bringing order and victory in Chechnya.' With emerging discursive agreement on what 'Russia' was and needed to be in order to fight off the existential terrorist threat, and with Putin at the helm of this re-defined and united 'Russia' bringing the fight to a victorious conclusion, he was authorized both to speak and act in a way that Yeltsin had not been for years.

Opinion polls reinforce the argument that Putin's position was bolstered by the process of securitization: from being virtually unknown to the Russian public when he was appointed Prime Minister, the rise in Putin's ratings is unprecedented among Russian politicians. According to polls conducted by the Russian Public Opinion Foundation, only 10 per cent of the Russian population held a positive opinion of Prime Minister Putin in the beginning of August 1999; by 4 September the figure had risen to 25 per cent.[13] But by 25 September it had reached 51 per cent. According to the Public Opinion Foundation, Putin's most appreciated characteristics were 'determination, endurance and decisiveness'.[14] In the 26 March 2000 presidential elections, Putin secured 53.4 per cent of the vote in the first round.

This is not the whole story of Putin's rise to power and the empowerment of the presidency in Russia since 1999, but it certainly indicates that discourses of danger should be studied as a vehicle for power, actor-hood and legitimacy in Russian politics. In focusing on the use of coercion to curb opposition in today's Russia, we must not forget how the legitimacy of the Putin regime came about in the first place, and how it has been sustained over time.

Intersubjectivity and responsibility

The Russian leadership and the securitizing narrative it promoted deserve attention in a study such as this. However, the most important insights on the securitization of Chechnya as a terrorist threat in Russia have probably been gained by broadening the focus beyond the current political leadership, in terms of time as well as space. The investigation of the Russian historical discursive terrain (Chapter 6) revealed 'Chechnya' or former versions of present-day 'Chechnya' repeatedly represented as different and dangerous: Russia's radical Other.

Thus, the official securitizing narrative of 1999 created its own, new content only to a certain extent. It drew heavily on a broader discursive foundation; several of the basic elements in the narrative already existed somewhere within this centuries-old debate. The discourse on Chechnya as an existential terrorist threat is therefore better understood as the updating of an already ingrained discursive structure that posited 'Chechnya' as different and dangerous and 'Russia' as a disciplining force. In trying to understand how the Second Chechen War became acceptable, this discursive fit between the official securitizing narrative and the Russian discursive terrain is significant. Since 'Chechnya' had been repeatedly invoked as a dangerous and different Other, it was easy to do so again.

That said, this book has deliberately sought to contradict the idea that discourses of radical Otherness in Russia are continuously reproduced in a mechanistic and uninterrupted way. To outline the genealogy of 'Chechnya' in Russia is not to claim that the Second Chechen War *had to* become a legitimate undertaking. A major part of this book has been devoted to studying representations of Chechnya and Russia in key 'audience groups' during autumn 1999. Even an official securitizing narrative that resonated well with the Russian discursive terrain *could* have been adjusted and even negated in statements and accounts by the political elite in the Russian Federal Assembly, by experts or journalists. The fact that the official narrative was received with resounding confirmation in their texts was not an inevitable outcome. The securitizing narrative could have been rejected, as it was in official statements during the interwar period. But that did not happen this time.

Chapter 7 detailed how an alternative position on Chechnya, one based on the interwar 'discourse of reconciliation', all but disappeared in the texts of the political elite in the Federal Assembly in the course of autumn 1999. Even statements by staunch liberal and democratic politicians such as Grigory Yavlinsky eventually slipped toward the dominant position on Chechnya as an existential terrorist threat, making violent retribution the only 'way out'. The vast majority of political elite texts not only endorsed the call for war implicit in official representations – they expanded on and detailed the gravity and omnipresence of the Chechen threat in such a way that they, as much as official statements, contributed to bringing the 'urgent security situation' into being. They also prescribed even more radical emergency measures than those indicated in the official narrative as the 'way out'. This means that it would be incorrect to view the political elite in Russia (beyond the presidency and government) as merely passive recipients of the official securitizing narrative. People with a position in or campaigning for a position in the chambers of the Russian Federal Assembly could have taken a clear stand against a new war – but few, if any, did so. Instead their statements contributed to the legitimization of the Second Chechen War.

Expert texts generally expanded on 'Chechnya' as a lawless and violent space, anchoring and authorizing this representation by historical references (Chapter 8). At times, characterizations in expert texts were as stark, emotional and terrifying as in political texts, employing descriptors of Chechnya that yielded a representation that can be placed at the top end of the existential threat scale. As a logical corollary to this threat representation, most expert texts argued that a violent and uncompromising approach to Chechnya was necessary, indeed even 'humanitarian'. Without wishing to place unreasonable weight on the argument, given the group of newspaper opinion pieces and editorials selected for use as sources of expert representations, I do want to stress the absence of balanced or tempered accounts of Chechen–Russian relations in expert language. While alternative positions could be identified, the majority of expert texts amounted to war-mongering on a par with official statements. In many ways this is a sad reflection of the difficulties of building independent and de-politicized expert communities in post-communist societies. If anything, the discursive mobilization for the Second Chechen War probably served to re-politicize what could have been a more independent expert community in Russia As for the potential role of the expert community in times of war, it is precisely when war is brewing that tempered, fact-based and balanced expert accounts *could* play a role. In autumn 1999, however, the Russian expert community was not willing or able to take on such responsibility for moderating the collective call for violent action.

Journalistic accounts proved to play a more significant role in the securitization of the Chechen threat than I had expected when I started work on this book. I saw the journalist community as a potential counterweight to Russian politicians, from where representations that negated and questioned the official securitizing narrative could emerge, as had been the case during the First Chechen War. However, my in-depth investigation of journalistic accounts, starting from the interwar period, has indicated the Russian media became a 'securitizing actor'. The accumulation of representations constructing Chechnya as an existential terrorist threat seemed to start on the pages of Russian newspapers.

Perhaps more than any other texts, the journalistic accounts reviewed in this book (before and after, with and without increasing government control of the media) contributed to the construction of Chechnya as an existential terrorist threat. As a crucial mediator of 'reality' to the broader Russian public, the Russian press contributed heavily to making the Second Chechen War acceptable. With a clear dichotomy created by merging everything Chechen into one category of 'dangerous' on the one hand, as opposed to a righteous and benevolent Russia united against the threat on the other, war must have appeared both logical and acceptable for those who relied on the major Russian newspapers.

While the material reviewed for this book did not explicitly include representations by the military, the police or the man in the street, indirect insights into the discourse prevalent in these groups also make clear the process of securitization as a broad and intersubjective endeavour. The 1999 securitizing narrative was not only fed from the 'sides' – i.e. from the texts of the political elite, the experts and the journalists; it was even fed from 'below'. Language on the micro-level reflected but also contributed to the fabric of the 'discourse of war'. Policemen in Moscow seemed to have their ways of speaking about Chechens and Caucasians. Although the statements referred to seemed an echo of the official narrative, they were probably more a re-articulation of an already-existing narrative, than a totally new position.

Turning to the soldiers and the generals, the juxtaposition between a dangerous and different Other and a righteous and united Russian Self is not surprising in the language of military men at war. Indeed, it could be said that such juxtapositions are necessary in any soldier's language, to make sense of why he should be shooting the guy on the 'other side of the river'. But such representations are seldom new inventions, nor can they be installed in a soldier in the course of a few days before sending him out to fight. It is more likely that the representation of the Chechen as a treacherous and dangerous Other was already fairly widespread in the language of Russian soldiers from the outset, and that these representations resonated with and merged with the new official representations of Chechnya as an existential terrorist threat. The Russian leadership told the police and the soldiers what kind of threat they were fighting, but this representation acquired both credibility and amplification because of representations that already existed among them.

The conceptualization of securitization as a fundamentally intersubjective process means that the positions of the 'securitizer' and the 'audience' are not set in stone. The representations in 'audience groups' discussed in this book are re-articulations of the official narrative, not carbon copies. They underscore the official narrative, but they also insert certain new aspects into the construction of the threat and authorize these contributions by the various features associated with these groups. The important point to take away here is how discourses of danger that dominate domestic constituencies below and beyond the official state level can feed into, shape and constrain official policy-positions over time. On the whole, the attention paid in this book to the wider Russian public and its important role via intersubjective discursive processes in the Russian polity supports the call by Ted Hopf[15] to give more consideration to Russian 'common sense' if we are to understand the choices and policies and the room for manoeuvre of the Russian political elites.

Returning to the Second Chechen War, we see that championing a broad and intersubjective understanding of securitization means that responsibility for legitimizing this violent undertaking cannot be confined to the political leadership. As the discourse on Chechnya as an existential terrorist threat became hegemonic, it subdued all other positions because it was confirmed in layer upon layer of texts voiced by groups looked upon merely as 'audience' in Copenhagen School securitization theory. This does not mean that responsibility evaporates, but that many more than a small clique at the top have to carry it. It also locates responsibility (and possibilities) for change at the doorstep of everyone with a voice. Although representations – the way we talk about ourselves and the Others – are normally part of an unreflective domain, people and communities can and should take responsibility for how they talk about others. This is particularly important when those Others are groups that already have a marginalized position in society or tend to be singled out for blame again and again.

The securitization of ethnic groups in a multi-ethnic state

What, then, of the consequences of the Second Chechen War for the Chechens as a group? I am not trying to say that the Second Chechen War was genocide. The official securitizing statements reviewed here indicate that the Russian leadership took care to invoke 'Chechnya' and not 'the Chechens' as the terrorist threat. However, such distinctions are not so easy to make or sustain in practice. With no explicit positive identity attached to the Chechens as a group, they readily became subsumed under the terrorist label as well. Once the war was under way, official statements also failed to differentiate the Chechen civilian population from the 'terrorist' threat in any explicit way. The consistent and many-faceted official securitizing narrative outlining and detailing the terrorist threat as an *existential* threat to Russia and the violent policies and practices necessary for dealing with it easily translated into an understanding of who *the Chechens* were, and what could and should be done to them.

The review of political elite, expert and journalistic texts also showed that representations by these groups often resulted in a construction of 'the Chechens' as 'terrorists'. Some even referred directly to 'the Chechens' as 'terrorists' or explained the collective guilt of 'the Chechens' in a way that official texts had not done. More significant perhaps is the way that 'Chechnya' and 'the Chechen fighter' were substantiated in layer upon layer of texts as different and dangerous and how this identification became attached to *all* Chechens. This was achieved through the constant reiteration of epithets combined with 'Chechen' ('Chechen bandits', 'Chechen terrorists', 'Chechen extremists'). Such slippage of identification from one object or group to another became particularly effective, as there existed no positive identifications that could give the Chechen population a human face. Despite the occasional acknowledgement of the existence of 'acceptable' pro-Russian 'Chechens' in some Russian journalistic accounts, the heavy construction of the 'suffering Chechen' which had put 'Chechens' on a par with 'Russians' as fellow human beings and which had dominated reporting during the First Chechen War was scarcely to be found in autumn 1999. Even as the war proceeded and potentially 'shocking events' were revealed, such as the bombing of civilian targets or atrocities against civilians, words presenting the Chechens as victims and Russia as guilty did not return to Russian newspaper pages. The Chechens were cast as instruments of terror. This served to legitimize violence against this group, both at the outset and as the war continued. It also served to mute one of the most potent mechanisms for mobilizing a population against war: feelings of identification and compassion with the target.

Beyond the war itself, the stigmatization of the Chechens as a group has implications for Russia as a multi-national, multi-confessional state. This study supports Erik Ringmar's[16] argument about war and how internal stability is created by excluding certain human collectives. But it also raises critical questions about the long-term consequences for stability when the human collective that is excluded resides within the same state and is expected to continue to do so in the future. At what cost was the internal cohesion in Russia generated by the Second Chechen War achieved?

Russia is and is bound to remain a multi-national and multi-confessional state. The 185 ethnic groups designated as 'nationalities' in the Russian Federation are intertwined geographically, economically and culturally, despite the existence of federal republics for the largest nationalities. Russia cannot afford to sustain internal cohesion by excluding and alienating groups on the basis of nationality or religion. The sharp divide between Chechens and Russians created by the wars is no less sharp today. Chechens quickly replaced the Jews as topping the lists of groups most disliked by Russians.[17] The radical alienation of Chechnya has not been ameliorated by the installation of the Kadyrov regime. As for the Chechens themselves, reliable opinion polls are hard to find. Let me instead cite the words of one Chechen lady who documented the killings of fellow villagers by Russian forces: 'After such hell, such impunity, such horror – who now could want to remain a part of Russia?'[18] The securitization of 'Chechnya' as an existential terrorist threat is not a case of articulating an identity of a territory/group as slightly different, thereby assigning them a slightly marginalized position in Russian society. No, this is a case of representing this group/territory as *radically* different and dangerous, and following up with massive and gross violence. This identity is not merely a linguistic statement: it is inscribed in Chechen bodies and Chechen lives.

Since the Second Chechen War, the internal divide between Russians and Chechens has broadened to encompass a wider group. I noted that Chechnya's neighbouring republics were included in the united Russian 'Self' at the beginning of the Second Chechen War. This is no longer the case. The eastern parts of the North Caucasus are dotted with armed militias that identify themselves as the righteous defenders of Salafi Islam, and see Russia as a different and dangerous infidel Other. These groups do have a certain appeal among the Muslim populations. According to official Russian sources, there are some 3,000 Russian Muslims, primarily from Northern Caucasus, fighting on the side of Islamic State in the Middle East. According to Russian researchers, the figure might be twice as large.[19]

From the Russian side, violence seems to be the only means of dealing with the challenge. There is a war-like situation in parts of the Northern Caucasus right now, and indiscriminate use of violence against the civilian population is not unusual. Counter-terrorist operations are carried out frequently, in Ingushetiya, Kabardino-Balkariya, Karchaevo-Cherkessia and, most intensely, in Dagestan.[20] Simultaneously, we have seen a heavy Russian bombing campaign in Syria conducted with the explicit argument that 'we must fight and eliminate them there, away from home'.[21] As Putin's 2015 speech referred to in the introductory chapter to this book illustrates, the hunt for 'terrorists', 'extremists' or 'Wahhabis' comes with new discursive mobilizations that outline these as existential threats to Russia. It is going to prove difficult to separate these broad designations from the Muslim populations that inhabit the Northern Caucasus.

How is Russia to deal with this critically fractured situation? Such deep, violized divides between groups that reside within the state seem to be easier to create than to mend. While today's Russian leadership appears to be falling back on policing and discipline as a means of bridging divides and governing the federation,[22] the long-term solution should be of a different kind. The articulation of a Russian identity that can encompass instead of alienate many different groups within the Russian Federation would be a first step – and a major challenge for the future. The second step should be a return to the Russian constitutional state, to secure the human rights for *all* citizens of the Russian Federation. On this account also the Second Chechen War has proven disastrous for Russia.

Violent practices, acceptance and the standing of human rights

Securitization theory holds that establishing an issue as an existential threat moves the issue out of the realm of normal politics and into the security realm allowing securitizing actors to claim 'a special right to use whatever means are necessary to block it'.[23] Whereas many applications and interpretations of securitization theory have focused on the 'special politics' implied in the first few words of that quote, I have deliberately paid more attention to the last words. This is not because 'special politics' are irrelevant in Russia. On the contrary, such politics seem to have become more the rule than the exception. The Second Chechen War marked the starting-point of a process that served to gather more and more power in the presidency, far beyond what is stipulated in the democratic constitution which Russia adopted in 1993.

Nevertheless, the primary focus in this book has been on how a discourse of existential threat makes it possible to use 'whatever means are necessary to block it'. This re-focusing was initially triggered by my awareness of the empirical case. I wanted to know how the very blunt means employed against Chechnya and Chechens during the Second Chechen War became acceptable.

Chapter 10 explored how the broad and many-layered classification of Chechnya as a terrorist threat immediately found material expression in the physical isolation of the Chechen Republic. Moreover, just as the classification of Chechnya easily slipped into a classification of 'the Chechens' as different and dangerous in linguistic representations, practices that served to 'seal off' Chechnya from Russia found a parallel in ways of sealing off the Chechens as such from Russia. Chechens were most logically placed in police stations or jails and dealt with by security personnel: after all, they could not be part of normal Russian society. The extremely intensive bombing of the tiny Chechen Republic ($17,300 \mathrm{km}^2$) from September 1999, which finally petered out early in 2001, seems absurd if one reviews the number of raids without taking in the discursive structures that make such violence possible. Taking into account the many-layered securitizing narrative which made 'Chechnya' stand out as an existential terrorist threat to Russia, however, these violent measures seem rather reasonable. Even the use of illegal weapons, indiscriminate bombing and border closures during bombing so that the civilian population could not escape could appear logical and legitimate against the background of the discursive mobilization in autumn 1999 (Chapter 11).

As to the ground offensive that followed the bombing, the 'cleansing operations' and the brute practices undertaken in connection with detentions at 'filtration points' entailed systematic, brutal and non-selective violence (Chapter 12). The Second Chechen War was as brutal as the First Chechen War had been, but this time brutality seemed to be called for and was seen as legitimate. This legitimacy was rooted in the identification of Chechnya as a 'huge terrorist camp' and in the official statements proclaiming that the 'toughest measures possible' had to be undertaken. At the same time, these physical practices of war contributed to constitute and reify Chechnya as an existential terrorist threat and Russia as a righteous defender.

Presumably, the public finds it easier to accept a call for war than the very concrete violence that a war entails. But statements by the political elite or commentary by the experts or the journalists on potentially shocking events examined in this study did not contain any protests against the violence that Chechnya and Chechens were subjected to during the war. Indeed, over the years, the use of emergency measures 'beyond rules that otherwise have to be obeyed' seemed to become self-evident and normal when carried out in Chechnya or against Chechens. The public opinion polls referred to in Chapter 9 also indicate that the broader Russian audience was willing to accept these practices of war as legitimate. I do not want to disregard the lack of information in Russia on what was going on in Chechnya when assessing how this acceptance was possible. Indeed, part of the reason why I wrote the chapters on practices of war in Chechnya was that such atrocities should be widely exposed and have not been in the Chechen case. What I wish to suggest is that categorizations such as those investigated throughout this book become crucial when something terrible occurs and there is uncertainty and lack of information on 'who did it'. Given the ingrained, repeated representations of Russia as innocent and Chechnya/Chechens as guilty and dangerous, responsibility for terrible deeds was most logically pinned on the Chechen side. It did not seem reasonable that Russian forces could have committed these atrocities – and if they did, that was probably necessary, given the existential threat that these people constituted.

Although I want to avoid sweeping claims, it is clear that the violent practices of the Second Chechen War have left today's Russia with a problematic heritage. There is undoubtedly a connection between the prevalence and legitimacy that these practices acquired in the beginning of the Second Chechen War and the acceptability of more private practices undertaken to 'cleanse' Russia of Chechens or so-called 'blacks' in later years. Memorial began its reporting of xenophobia from June 2003 'because mounting xenophobia was obvious in all spheres of life and negatively affected those of the people from Chechnya who lived outside their republic … people of obviously non-Slavic extraction are more and more frequently attacked in the streets'.[24] Moving to the North Caucasian theatre again, Russia's recent efforts to curb the growing local insurgency in Dagestan include bombing of the territory and *zachistki* of the villages, as well as a general militarization of the republic.[25] The *zachistki* and bombing practices distinctly resemble those carried out in Chechen villages at the beginning of this century.[26]

Just as problematic as the social acceptance of such violent and indiscriminate practices against citizens of the Russian Federation is the weakening of the legal foundation of fundamental human rights in Russia in the wake of the Second Chechen War. A whole series of key provisions in the Russian Constitution as well as in international legal regimes to which Russia is committed were abrogated or challenged in the heat of securitization of Chechnya as a terrorist threat. I have detailed how the practices reviewed amounted to discrimination on the basis of nationality, and undermined the freedom of movement and the freedom of expression. Most fundamentally, the right to life and the right to freedom from torture and inhumane treatment were violated. There were also serious violations of Russia's obligations under the Geneva Conventions of 1949 and Protocol II to the Convention which specifies the rules for internal armed conflict, and under the instruments of international human rights law to which Russia is also party.

That such massive and gross violations of fundamental human rights contribute to undermine the *spirit* of the Constitution is beyond doubt, particularly when the Russian authorities have been in a state of denial over abuses. Impunity for perpetrators of gross human rights violations and war crimes has been the rule, not the exception. This book has also documented how several new laws and orders that were in breach of the Russian Constitution (or existing Russian legislation) were adopted and endorsed in the heat of the fight against the existential terrorist threat.

With the benefit of hindsight, it seems as if these laws marked the beginning of a much broader process of constitutional backsliding. The introduction and endorsement of the law on restricting media appearances by members of the armed formations and punishing the outlets that publish such articles, for example, was the first in a row of measures and laws that over time have resulted in the decline of media freedom in Russia.[27] Likewise, Prime Minister Putin's statement that the new war did *not* need a strong legal foundation has not been without repercussions. The acceptance of undertaking military operations against citizens of the Russian Federation based on the 1998 'Law on Combating Terrorism', which implied that the relation between forces and civilian population are not regulated by law during these operations, has become even stronger over time. This 'black box' situation of unregulated relations between forces and civilian population has now become the standard pattern when Russian forces deal with the unrest which has spread to several republics in North Caucasus. The practice of using Russian military forces in counter-terrorist operations against Russian citizens even became so accepted and normalized that it was codified: the 2006 law 'On counteraction of terrorism' which replaced the 1998 law 'On combating terrorism' legalized this practice.[28]

Although it would be wrong to exaggerate the impact that the moral and formal acceptance of emergency measures during the Second Chechen War has had on the course of Russia's development towards a polity based on the rule of law and rights, it must not be underestimated either. Even societies with more ingrained law and rights-based traditions set these aside in times of war. Russia is in many ways a newcomer on this account, with the years since 1991 constituting a potential new beginning. Russia's short history as a law and rights-based society has probably rendered it particularly vulnerable to strong securitizations, making it easier to agree within the political elite on the moves 'beyond rules that otherwise have to be obeyed' and taking Russia further away from the clear human rights foundation codified in the 1993 Constitution.

The life of Aslan Maskhadov as a microcosm of the Second Chechen War

Representations of the Chechen President Aslan Maskhadov as an 'event within the event' have been reviewed throughout this book. In many ways, his trajectory follows the trajectory of 'Chechnya', from being a de-securitized issue in the interwar period to becoming a highly securitized issue necessitating violent retribution from 1999 onward. Maskhadov was moved from the position of an equal and reliable partner to one of radical difference and danger, a position that warranted violent death.

In the interwar period, Russian official representations of Maskhadov depicted him as a legitimately elected president, a reliable partner and a guarantor of stability, the rule of law and human rights in the region. The meaning attached to 'Maskhadov' in official statements in summer 1999 was not a direct equation with 'terrorism' – but there were changes compared to the official interwar representations. No longer represented as a victim, he was now portrayed as consenting to terrorism, through the expression that it 'suits him'. The policies proposed in October for making the Chechen Parliament the 'only legitimate organ of power in Chechnya' – in effect rendering President Maskhadov 'illegitimate' – were logical, in view of the new identity he was given in official discourse.

But the old version of 'Maskhadov' still lingered on in some representations. In the texts of the political elite, we saw an alternative position that clearly distinguished 'Maskhadov' from the terrorist threat, and portrayed him as a legitimate and reliable partner. Such an identity construction made policies of negotiation and cooperation seem logical and legitimate. And indeed, calls by liberal Russian politicians such as Grigory Yavlinsky and even members of the CPRF for negotiating with Maskhadov were noted in October and November 1999, and appeared again in 2000. Nevertheless, the defence of the old version of 'Maskhadov' became increasingly half-hearted. In the language of the majority of the political elite, 'Maskhadov' was gradually shifted from being a legitimate and trustworthy partner to being unreliable and weak, potentially an accomplice of the terrorists. A similar pattern emerged in expert texts. In journalistic accounts, 'Maskhadov' seemed to disappear altogether. He was not given a distinct position: he was simply subsumed under the terrorist threat.

All the same, Maskhadov's status remained contested in the years that followed 1999. His name kept coming up whenever there was talk of the need for negotiations, and images of Maskhadov as someone one could and should talk to lingered on.[29] Over time, however, his position became more and more precarious in official representations. With increasing official control over the media in Russia, it was repeatedly announced that video material confirming Maskhadov's personal instruction of terrorists had been found.[30] Finally, the repeated 'documentation' by the FSB of Maskhadov as a terrorist somehow served to decide the discursive struggle over who 'Maskhadov' was. In particular, the video documentation that purported to implicate him in terrorism presented after the terrorist attack at the Nord-Ost theatre in Moscow in 2002 served to place 'Maskhadov' squarely on the side of the existential terrorist threat.[31] With this, the prospects of a negotiated solution to the Chechen conflict which admittedly had been meagre for a long time, plummeted to zero. There was no one to negotiate with among the Chechens fighting against 'Russia' in Chechnya. There could be no negotiations.[32]

For Maskhadov personally, the consequences were fatal. A $10 million reward was placed on his head. When army spokesman Ilya Shabalkin informed the press that Maskhadov had been killed, he said: 'The Federal Security Forces, while conducting a special operation … killed international terrorist and rebel leader Aslan Maskhadov.'[33] Even in death, Maskhadov's life was no longer understood as human, but as 'terrorist'. In accordance with Order No. 164 of 20 March 2003, his body was not handed over for Muslim burial. It was interred in an unmarked terrorist grave, the location not communicated to his family and relatives.

Notes

1 S. Guzzini. 2011. Securitization as a causal mechanism. *Security Dialogue*, 42(4/5): 329–341, 334.
2 For reliable figures on troops and casualties in the North Caucasus, see *Caucasian Knot* available at http://eng.kavkaz-uzel.ru/ and accessed 19 February 2016.
3 I.B. Neumann. 2010. National security, culture and identity. In: *The Routledge Handbook of Security Studies*, edited by V. Mauer and M.D. Cavelty. New York: Routledge, 95.
4 W.E. Connolly. 1991. *Identity/Difference: Democratic negotiations of political paradox*. Ithaca, NY: Cornell University Press.
5 See, for example, S. Medvedev. 2004. 'Juicy morsels': Putin's Beslan address and the construction of the new Russian identity. *Ponars Policy Memo* (334); E.B. Kolt and C. Wallander (eds.) 2007. *Russia Watch: Essays in honor of George Kolt*. Washington, DC: Center for Strategic and International Studies.
6 See, for example, Putin's speech to the Valdai International Discussion Club on 19 September 2013, where he underlined the fundamental importance of finding and strengthening a national identity for Russia and then set out to identify what values this national identity should comprise. He indicated sovereignty, independence and territorial integrity as key markers for Russia as a state, while also stressing patriotism, traditional spiritual and moral values as well as multi-culturalism and multi-ethnicity as fundamental Russian values (available at http://eng.kremlin.ru/news/6007 and accessed 17 February 2016). The speech President Putin made on 18 March 2014 in connection with the annexation of Crimea articulated a more ethnic nationalist Russian identity (available at http://en.kremlin.ru/events/president/news/20603 and accessed 17 February 2016).
7 For an alternative and distinctly empirical account of how the Second Chechen War contributed to Putin's rise to power, see J. Headley. 2005. War on Terror or pretext for power? Putin, Chechnya, and the 'Terrorist International'. *Australian Journal of Human Security*, 1 (2): 13–35.
8 'Putin sozdal klub prem'yer-ministrov', *NeGa*, 6 October 1999.
9 'Putin predlagayet novyy plan chechenskogo uregulirovaniya', *NeGa*, 15 September 1999.
10 'Vybory na fone teraktov', *NG Regiony*, 28 September 1999.
11 'Demokraticheskiy glava pravitel'stva ili 'voyennyy prem'yer'?', *NeGa*, 14 October 1999.
12 See for example 'Arestovan Nadir Khachilayev', *NeGa*, 8 October 1999, 'Po svoim iz-za ugla', *RoGa*, 20 October 1999, 'Federal'nyye voyska podoshli k Groznomu', *NeGa*, 21 October 1999.
13 'Putin zavoyëvyvayet simpatii rossiyan', *NeGa*, 11 September 1999.
14 B.S. Doktorov, A.A. Oslon and E.C. Petrenko. 2002. *Epokha Yeltsina: Mneniya Rossiyan*. Moskva: Institut Fonda Obshchestvennoye Mneniye, 308–309.

15 T. Hopf. 2013. Common-sense constructivism and hegemony in world politics, *International Organization*, 67(2): 317–354.

16 E. Ringmar. 1996. *Identity, Interest and Action. A cultural explanation of Sweden's intervention in the Thirty Years War.* Cambridge: Cambridge University Press.

17 For a collection of statistics that illustrate the sharp divide between Chechens/North Caucasians and Russians today, see the 2013 report by the Valdai Club available at http://vid-1.rian.ru/ig/valdai/Russian_Identity_2013_rus.pdf and accessed 19 February 2016.

18 Kheda Muskhadzhiyeva, quoted in interview with Maura Reynolds, *Los Angeles Times* staff writer reporting from Nazran on 24 April 2001, as carried on *Johnson's Russia List* 25 April 2001.

19 Conversation with Akhmed Yarlikapov in September 2015, senior research fellow at the Center for Caucasian Studies and Regional Security, MGIMO University.

20 For reports on the unfolding events in North Caucasus, see Jamestown Foundation's *Eurasia Daily Monitor*, available at www.jamestown.org/programs/edm/ and *North Caucasus Weekly*, available at www.jamestown.org/nc/ or daily news bulletins on *Caucasian Knot*, available at http://northcaucasus.eng.kavkaz-uzel.ru/ and accessed 22 February 2016.

21 Presidential Address to the Federal Assembly, 3 December 2015, available at Kremlin.ru (http://en.kremlin.ru/events/president/news/50864) and accessed 22 February 2016.

22 I.B. Neumann. 2011. Governing a great power: Russia's oddness reconsidered. In: *Governing the Global Polity: Practice, mentality, rationality*, edited by I. Neumann and O.J Sending. Ann Arbor, MI: University of Michigan Press, 70–110.

23 O. Wæver. 1995a. Securitization and desecuritization. In: R.D. Lipschutz (ed.) *On Security*. New York: Columbia University Press, 55.

24 Cited in Svetlana Gannushkina (ed.) (2005) *On the Situation of the Residents of Chechnya in the Russian Federation*, available at www.memo.ru/2009/05/26/2605092.htm and accessed 22 February 2016.

25 Mairbek Vatchagaev. 2016. 'Russian Defense Ministry Holds More Exercises in Dagestan' *North Caucasus Weekly* available at www.jamestown.org/programs/nc/single/?tx_ttnews%5Btt_news%5D=45089&tx_ttnews%5BbackPid%5D=24&cHash=2527b53e9fd1ab6c4994512421bcf4cb and accessed 22 February 2016.

26 Emil Souleimanov. 2013. Mopping Up Gimry: "Zachistkas" Reach Dagestan *Central Asia-Caucasus Analyst*, 24 April 2013, available at www.cacianalyst.org/publications/analytical-articles/item/12706-mopping-up-gimry-zachistkas-reach-dagestan.html, and accessed 22 February 2016. On the bombings in Dagestan see Emil Souleimanov. 2012. Russia re-deploys troops to Dagestan, *Central Asia-Caucasus Analyst*, 14 November 2012, available at http://old.cacianalyst.org/?q=node/5878 and accessed 22 February 2016.

27 On this, see J. Wilhelmsen. 2003. *Norms: The forgotten factor in Russian–Western rapprochement: a case study of freedom of the press under Putin.* FFI Report 00457. Kjeller: FFI; M. Lipman and M.A. McFaul. 2005. Putin and the media. In: D.R. Herspring (ed.) *Putin's Russia: Past imperfect, future uncertain.* Lanham, MD: Rowman and Littlefield, 63–84; S. Gehlbach. 2010. Reflections on Putin and the media. *Post-Soviet Affairs*, 26(1): 77–87.

28 M. Omelicheva. 2009. Russia's counterterrorism legislation, warts and all. In: A. Mikkonen (ed.) *Threats and Prospects in Combating Terrorism: Series 2, Research Reports No 41.* Helsinki: National Defence University, Department of Strategic and Defence Studies, 1–20: 4–9.

29 Indeed, the first discussions between the adversaries since the beginning of the war were held in November 2001 between Maskhadov's representative Akhmed Zakayev and Russia's Representative to the Southern Federal District Viktor Kazantsev. First Deputy Chief of the General Staff of the Russian Federation General Manilov stated in December 1999 that he saw Maskhadov as a 'legitimate figure' although he wa

'almost totally in the power of gangsters and terrorists' ('Chechen President: a Legitimate Figure', *Itar Tass*, 16 December 1999).
30 OSCE. 2003. Representative on freedom of the media. In: *Freedom and Responsibility Yearbook 2002/2003*. Vienna: OSCE, 252.
31 'Khasavyurta ne budet', *Izvestiya*, 11 November 2002. See also 'Ne ostalos' somneniy v prichastnosti Maskhadova k teraktu v Moskve', *RoGa*, 4 February 2003.
32 On the construction of Maskhadov's illegitimacy, see Aurèlie Campana and Kathia Légaré (2010, 52–54).
33 'Chechen leader Maskhadov is killed, Army reports', *Reuters*, 8 March 2005.

Bibliography

Adler, E. and Pouliot, V. 2011. *International Practices*. Cambridge: Cambridge University Press.

Amnesty International UK. 2000a. *Rape and Torture of Children in Chernokozovo 'filtration Camp'*. Available at: www.amnesty.org.uk/news_details.asp?NewsID=12978. [Accessed 31 July 2013].

Amnesty International. 2000b. *Real Scale of Atrocities in Chechnya: New evidence of cover-up*. Available at: www.amnesty.org/fr/library/asset/EUR46/020/2000/en/3a33f6e5-3e4d-407f-b3c6-9f1b60ce584a/eur460202000en.pdf. [Accessed 31 July 2013].

Amnesty International. 2002. *Denial of Justice*. London: Amnesty International Publications. Available at: www.amnesty.org/en/library/info/EUR46/027/2002. [Accessed 31 July 2013].

Amnesty International. 2003. *Rough Justice: The law and human rights in the Russian Federation*. London: Amnesty International Publications. Available at: www.amnesty.org/en/library/info/EUR46/054/2003/en. [Accessed 31 July 2013].

Aradau, C. 2004. Security and the democratic scene: Desecuritization and emancipation. *Journal of International Relations and Development*, 7(4), 388–413.

Austin, J.L. 1962. *How to Do Things with Words*. Oxford: Clarendon Press.

Åtland, K. and Ven Bruusgaard, K. 2009. When security speech acts misfire: Russia and the Elektron incident. *Security Dialogue*, 40(3), 333–353.

Bacon, E. and Renz B., with Cooper J. 2006. *Securitizing Russia: The domestic politics of Putin*. Manchester: Manchester University Press.

Baddeley, J.F. 1908. *The Russian Conquest of the Caucasus*. London: Longmans, Green.

Balzacq, T. 2005. The three faces of securitization: Political agency, audience, and context. *European Journal of International Relations*, 11(2), 171–201.

Balzacq, T. (ed.) 2011. *Securitization Theory: How security problems emerge and dissolve*. Abingdon: Routledge.

Barth, F. (ed.) 1969. *Ethnic Groups and Boundaries*. Oslo: Norwegian University Press.

Bigo, D. 2002. Security and immigration: Towards a critique of the Governmentality of Unease. *Alternatives*, 27(1), 63–92.

Blandy, C.W. 2000. *Chechnya: Two federal interventions. An interim comparison and assessment*. Camberley, UK: Conflict Studies Research Centre, Royal Military Academy, Sandhurst.

Brecher, B., Devenney, M. and Winter, A. (eds.) 2010. *Discourses and Practices of Terrorism: Interrogating terror*. London: Routledge.

Browning, C.S. and McDonald, M. 2013. The future of critical security studies: Ethics and the politics of security. *European Journal of International Relations*, 19(2), 35–255.

Bullough, O. 2010. *Let our Fame Be Great: Journeys among the defiant people of the Caucasus*. London: Penguin.

Butler, J. 1993. *Bodies that Matter: On the discursive limits of sex*. London: Routledge.

Buzan, B. 1997. Rethinking security after the Cold War. *Cooperation and Conflict*, 32(1), 5–28.

Buzan, B. and Hansen, L. 2009. *The Evolution of International Security Studies*. Cambridge: Cambridge University Press.

Buzan, B. and Wæver, O. 1997. Slippery? Contradictory? Sociologically untenable? The Copenhagen School replies. *Review of International Studies*, 23(2), 171–201.

Buzan, B. and Wæver, O. 2003. *Regions and Powers*. Cambridge: Cambridge University Press.

Buzan, B. and Wæver, O. 2009. Macrosecuritization and security constellations: reconsidering scale in securitization theory. *Review of International Studies*, 35(2), 253–276.

Buzan, B., Wæver, O. and de Wilde, J. 1998. *Security: A new framework for analysis*. Boulder, CO: Lynne Rienner.

Campana, A. and Légaré, K. 2010. Russia's counterterrorism operation in Chechnya: Institutional competition and issue frames. *Studies in Conflict and Terrorism*, 43(1), 47–63.

Campbell, D. 1992. *Writing Security*. Minneapolis, MN: University of Minnesota Press.

C.A.S.E. Collective. 2006. Critical approaches to security in Europe. A networked manifesto. *Security Dialogue*, 37(4), 443–487.

Cherkasov, A. and Grushkin, D. 2005. The Chechen wars and the struggle for human rights. In: R. Sakwa (ed.) *Chechnya: From past to future*. London: Anthem Press, 131–157.

Closs, S., Vaughan-Williams, A. and Vaughan-Williams, N. 2010. *Terrorism and the Politics of Response*. London: Routledge.

Connolly, W.E. 1991. *Identity/Difference: Democratic negotiations of political paradox*. Ithaca, NY: Cornell University Press.

Croft, S. 2006. *Culture Crisis and America's War on Terror*. Cambridge: Cambridge University Press.

Croft, S. 2012. *Securitizing Islam: Identity and the search for security*. Cambridge: Cambridge University Press.

Dalby, S. 1988. Geopolitical discourse: The Soviet Union as Other. *Alternatives*, 13(4), 141–155.

Dannreuther, R. 2010. Islamic radicalization in Russia: An assessment. *International Affairs*, 86(1), 109–126.

Der Derian, J. 2005. Imagining terror: Logos, Pathos and Ethos, *Third World Quarterly*, 26(1), 23–37.

Derrida, J. 1967. *Of Grammatology*. Baltimore: The Johns Hopkins University Press.

Dillon, Michael. 1990. The alliance of security and subjectivity. *Current Research on Peace and Violence*, 13(3), 101–124.

Derrida, J. 1981. *Positions*. Chicago, IL: University of Chicago Press.

Doktorov, B.S., Oslon, A.A. and Petrenko, E.C. 2002. *Epokha Yeltsina: Mneniya Rossiyan*. Moskva: Institut Fonda Obshchestvennoye Mneniye.

Donnelly, F. 2013. *Securitization and the Iraq War: The rules of engagement*. New York: Routledge.

Doty, R. 1996. *Imperial Encounters*. Minneapolis, MN: University of Minnesota Press.

Doty, R. 1997. Aporia: A critical exploration of the agent–structure problematique in international relations theory. *European Journal of International Relations*, 3(3), 365–392.

Doty, R. 1998–1999. Immigration and the politics of security. *Security Studies* 8(2–3), 71–93.

Dunlop, J.B. 1998. *Russia Confronts Chechnya: Roots of a separatist conflict.* Cambridge: Cambridge University Press.

Evangelista, M. 2002. *The Chechen Wars. Will Russia go the way of the Soviet Union?* Washington, DC: Brookings Institution Press.

Fierke, K. 2004. World or worlds? The analysis of content and discourse. *Qualitative Methods: Newsletter of the American Political Science Association Organized Section on Qualitative Methods*, 2(1), 36–39.

Fierke, K.M. 2007. *Critical Approaches to International Security.* Cambridge: Polity.

Floyd, R. 2010. *Security and the Environment: Securitization theory and US environmental security policy.* Cambridge: Cambridge University Press.

Foucault, M. 1972. *The Archaeology of Knowledge and the Discourse on Language.* Pantheon Books: New York.

Foucault, M. 1995. *Discipline and Punish: The birth of the prison.* New York: Vintage Books.

Galeotti, M. 2010. *The Politics of Security in Modern Russia.* Farnham, UK: Ashgate.

Gall, C. and Waal, T.D. 1997. *Chechnya: A small victorious war.* London: Pan Original.

Gammer, M. 2006. *The Lone Wolf and the Bear. Three centuries of Chechen defiance of Russian rule.* London: Hurst.

Gannushkina, S. (ed.) 2005. *On the Situation of the Residents of Chechnya in the Russian Federation.* Moscow: Memorial Human Rights Center, 'Migration Rights' network. Available at: www.memo.ru/2009/05/26/2605092.htm. [Accessed 31 July 2013].

Gehlbach, S. 2010. Reflections on Putin and the media. *Post-Soviet Affairs*, 26(1), 77–87.

Gerber, T.P. and Mendelson, S.E. 2002. Russian public opinion on human rights and the war in Chechnya. *Post-Soviet Affairs*, 18(4), 271–305.

Gevorkjan, N., Timakova, N. and Kolesnikov, A. 2000. *First Person: An astonishingly frank self-portrait by Russia's president.* New York: Public Affairs.

Gilligan, E. 2010. *Terror in Chechnya. Russia and the tragedy of civilians in war.* Princeton, NJ: Princeton University Press.

Grayson, K. 2003. Securitization and the boomerang debate: A rejoinder to Liotta and Smith-Windsor. *Security Dialogue*, 34(3), 337–343.

Guzzini, S. 2011. Securitization as a causal mechanism. *Security Dialogue*, 42(4/5), 329–341.

Haas, M. 2003. *The Use of Russian Airpower in the Second Chechen War.* Swindon: Defence Academy of the United Kingdom, Conflict Studies Research Centre.

Hacking, I. 1999. *The Social Construction of What?* Cambridge, MA: Harvard University Press.

Hagmann, J. 2015. *(In-)Security and the Production of International Relations: The politics of securitization in Europe.* London: Routledge.

Hale, H.E. 2000. The parade of sovereignties: Testing theories of secession in the Soviet setting. *British Journal of Political Science*, 30(1), 31–56.

Hale, H.E. 2004. Yabloko and the challenge of building a liberal party in Russia. *Europe-Asia Studies*, 56(7), 993–1020.

Hansen, L. 2000. The little mermaid's silent security dilemma and the absence of gender in the Copenhagen School. *Millennium*, 29(2), 289–306.

Hansen, L. 2006. *Security as Practice.* New York: Routledge.

Hansen, L. 2011. The politics of securitization and the Mohammad cartoon crises: A post structuralist perspective. *Security Dialogue*, 42(4/5), 357–369.

Hardy, C., Harley, B. and Phillips, N. 2004. Discourse analysis and content analysis: Two solitudes. *Qualitative Methods: Newsletter of the American Political Science Association Organized Section on Qualitative Methods*, 2(1), 19–22.

Hayes, J. 2009. Identity and securitization in the democratic peace: The United States and the divergence of response to India and Iran's nuclear programs. *International Studies Quarterly*, 53(4), 977–999.

Headley, J. 2005. War on Terror or pretext for power? Putin, Chechnya, and the 'Terrorist International'. *Australasian Journal of Human Security*, 1 (2), 13–35.

Holland, J. 2013. *Selling the War on Terror: Foreign policy discourses after 9/11*. New York: Routledge.

Holzscheiter, A. 2010. *Children's Rights in International Politics: The transformative power of discourse*. London: Palgrave Macmillan.

Hopf, T. 2013. Common-sense constructivism and hegemony in world politics. *International Organization*, 67(2), 317–354.

Houen, A. (ed.) 2014. *States of War since 9/11: Terrorism, sovereignty and the War on Terror*. London: Routledge.

Hughes, J. 2007. *Chechnya: From nationalism to Jihad*. Philadelphia, PA: University of Pennsylvania Press.

Human Rights Watch. 2000a. *Russia/Chechnya – February 5: A day of slaughter in Novye Aldi*. Available at: www.hrw.org/reports/2000/06/01/russiachechnya-february-5-day-slaughter-novye-aldi. [Accessed 31 July 2013].

Human Rights Watch. 2000b. *Welcome to Hell*. Available at: www.refworld.org/docid/3ae6a8750.html. [Accessed 8 October 2013].

Human Rights Watch. 2000c. *Hundreds of Chechens Detained in 'Filtration Camps'. Detainees face torture, extortion, rape*. Available at: www.hrw.org/news/2000/02/17/hundreds-chechens-detained-filtration-camps. [Accessed 31 July 2013].

Human Rights Watch. 2000d. *No Happiness Remains. Civilian killings, looting, and rape in Alkhan-Yurt, Chechnya*. Moscow: Memorial. Available at: www.hrw.org/reports/2000/russia_chechnya2/index.htm. [Accessed 31 July 2013].

Human Rights Watch. 2000e. *Civilian Killings in Staropromyslovsky District of Grozny*. Available at: www.hrw.org/reports/2000/russia_chechnya/. [Accessed 31 July 2013].

Human Rights Watch. 2001. *The Dirty War in Chechnya: Forced disappearances, torture and summary executions*. Available at: www.hrw.org/reports/2001/chechnya/RSCH0301.PDF. [Accessed 31 July 2013].

Human Rights Watch. 2002. *Last seen ...: Continued 'disappearances' in Chechnya*. Available at: www.hrw.org/reports/2002/04/15/last-seen-0. [Accessed 31 July 2013].

Huysmans, J. 1996. Migrants as a security problem: Dangers of 'securitising' societal issues. In: R. Miles and D. Thranhardt (eds.) *Migration and European Integration*. London: Pinter, 53–72.

Huysmans, J. 1998. Revisiting Copenhagen: Or, on the creative development of a security studies agenda in Europe. *European Journal of International Relations*, 4(4), 479–505.

Huysmans, J. 1999. The question of the limit: Desecuritisation and the aesthetics of horror in political realism. *Millennium*, 27(3), 569–590.

Ingram, A. 1999. A nation split into fragments: The Congress of Russian Communities and Russian nationalist ideology. *Europe–Asia Studies*, 51(4), 687–704.

Jackson, P. 2006. *Civilizing the Enemy: German reconstruction and the invention of the West*. Ann Arbor, MI: University of Michigan Press.

Jackson, P.T. 2011. *The Conduct of Inquiry in International Relations*. New York: Routledge.

Jackson, R. 2005. *Writing the War on Terrorism: Language, politics and counter-terrorism.* Manchester: Manchester University Press.

Jackson, R., Murphy, E. and Poynting, S. (eds.) 2011. *Contemporary State Terrorism: Theory and practice.* London: Routledge.

Jackson, R., Smyth, M.B. and Gunning, J. (eds.) 2009. *Critical Terrorism Studies: A new research agenda.* London: Routledge.

Jørgensen, M. and Phillips, L. 2002. *Discourse Analysis as Theory and Method.* London: Sage.

Kassianova, A. 2001. Russia: Still open to the West? Evolution of the state identity in the foreign policy and security discourse. *Europe–Asia Studies*, 53(6), 821–839.

King, C. and Menon, R. 2010. Prisoners of the Caucasus. *Foreign Affairs* 89(4), 20–35.

Kline, E. 1995. *Chechen History.* [online] Available at: www.newsbee.net/moscow/chhistory.html. [Accessed 26 July 2013].

Kolt, E.B. and Wallander, C. (eds.) 2007. *Russia Watch: Essays in honor of George Kolt.* Washington, DC: Center for Strategic and International Studies.

Krause, K. and Williams, M.C. 1997. From strategy to security: Foundations of critical security studies. In: K. Krause and M.C. Williams (eds.) *Critical Security Studies.* London: Routledge, 33–61.

Krebs, R.R. and Jackson, P.T. 2007. Twisting tongues and twisting arms: The power of political rhetoric. *European Journal of International Relations*, 13(1), 35–66.

Krebs, R.R. and Lobasz, J.K. 2007. Fixing the meaning of 9/11: Hegemony, coercion, and the road to war in Iraq. *Security Studies* 16 (3), 409–451.

Kristeva, J. 1980. *Desire in Language: A semiotic approach to literature and art.* New York: Columbia University Press.

Laclau, E. and Mouffe, C. 1985. *Hegemony and Socialist Strategy: Towards a radical democratic politics.* London: Verso.

Lermontov, M. 1977. Cossack lullaby. In: L. Kelly (trans.) *Tragedy in the Caucasus.* London: Constable.

Levashov, B.K. 2001. *Rossiyskoye obshchestvo i radikal'nye reformy.* Moscow: Akademia, Russian Academy of Science, Institute of Social-political Research.

Lieven, A. 1998. *Chechnya: Tombstone of Russian power.* New Haven, CT: Yale University Press.

Lipman, M. and McFaul, M.A. 2005. Putin and the media. In: D.R. Herspring (ed.) *Putin's Russia: Past imperfect, future uncertain.* Lanham, MD: Rowman and Littlefield, 63–84.

Mäkinen, S. 2008. Russian geopolitical visions and argumentation. Parties of power, democratic and communist opposition on Chechnia and NATO, 1994–2003. In: *Acta Universitatis Tamperensis* 1293. Tampere: Tampere University Press.

Malinova O. 2012. Russia and 'the West' in the twentieth century: A binary model of Russian culture and transformations of the discourse on collective identity. In: *Constructing Identities in Europe: German and Russian perspectives.* Baden-Baden: Nomos Verlagsgesellschaft, 63–79.

March, L. 2001. For victory? The crises and dilemmas of the Communist Party of the Russian Federation. *Europe–Asia Studies*, 53(2), 263–290.

McDonald, M. 2008. Securitization and the construction of security. *European Journal of International Relations*, 14(4), 563–587.

McSweeney, B. 1996. Identity and security: Buzan and the Copenhagen School. *Review of International Studies*, 22(1), 81–93.

Medvedev, S. 2004. 'Juicy morsels': Putin's Beslan address and the construction of the new Russian identity. *Ponars Policy Memo* (334).

Meier, A. 2005. *Chechnya. To the heart of a conflict*. New York: W.W. Norton.

Memorial. 1999a. *Moscow after the Explosions. Ethnic cleansings. September–October 1999.* Available at: www.memo.ru/eng/hr/ethn-e2.html. [Accessed 31 July 2013].

Memorial 1999b. *Ethnic Discrimination and Discrimination on the Basis of Place of Residence in the Moscow Region.* Available at: www.memo.ru/eng/hr/ethn-el.html. [Accessed 31 July 2013].

Memorial. 1999c. *The Missile Bombing of Grozny, October 21, 1999.* Available at www. memo.ru/eng/memhrc/texts/bom.shtml. [Accessed 18 March 2016].

Memorial. 1999d. *Point Strokes.* Available at: www.memo.ru/eng/memhrc/texts/bom. shtml. [Accessed 18 March 2016].

Memorial. 2000. *Compliance of the Russian Federation with the Convention on the Elimination of all Forms of Racial Discrimination.* Available at: www.memo.ru/hr/discrim/ ethnic/disce00.htm. [Accessed 18 March 2016].

Memorial. 2001. *The 'Cleansing' Operations in Sernovodsk and Assinovskaya were Punishment Operations.* Available at www.memo.ru/eng/memhrc/texts/sern&assin.shtml. [Accessed 18 March 2016].

Memorial. 2002a. *Myths and Truth about Tsotsin-Yurt December 30, 2001–January 3, 2002.* Available at www.memo.ru/eng/memhrc/texts/mythtruth.shtml. [Accessed 18 March 2016].

Memorial. 2002b. *Swept under: Torture, forced disappearance, and extrajudicial killings during sweep operations in Chechnya.* Available at: www.hrw.org/reports/2002/russ-chech/chech0202.pdf. [Accessed 18 March 2016].

Memorial. 2008a. *Special Operations [zachistka].* Available at www.memo. ru/2008/09/04/0409081eng/part5.htm. [Accessed 18 March 2016].

Memorial. 2008b. *Filtration System.* Available at: www.memo.ru/2008/09/04/0409081eng/ part61.htm. [Accessed 18 March 2016].

Memorial and Civic Assistance. 1999. *The Report on the Observer Mission to the Zone of the Armed Conflict, Based on the Inspection Results in Ingushetia and Chechnya.* Available at: www.memo.ru/eng/memhrc/texts/ch2.shtml. [Accessed 18 March 2016].

Memorial and Demos. 2007. *Counterterrorism Operation by the Russian Federation in the Northern Caucasus throughout 1999–2006.* Available at: www.memo.ru/hr/ hotpoints/N-Caucas/dkeng.htm. [Accessed 18 March 2016].

Mikiewicz, E. 1997. *Changing Channels: Television and the struggle for power in Russia.* New York: Oxford University Press.

Mikiewicz, E. 2008. *Television, Power, and the Public in Russia.* Cambridge: Cambridge University Press.

Millken, J. 1999. The study of discourse in international relations: A critique of research and methods. *European Journal of International Relations*, 5(2), 225–254.

Mintzen, J. 2006. Ontological security in world politics: State identity and the security dilemma. *European Journal of International Relations*, 12(3), 341–370.

Moore, C. and Tumelty, P. 2009. Unholy alliances in Chechnya: From Communism and nationalism to Islamism and Salafism. *Journal of Communist Studies and Transition Politics*, 25(1), 73–94.

Mutimer, D. 2007. Critical security studies: A schismatic history. In: A. Collins (ed.) *Contemporary Security Studies.* Oxford: Oxford University Press, 53–75.

Møller, F. 2007. Photographic interventions in post-9/11 security policy. *Security Dialogue*, 38(2), 179–96

Neumann, I.B. 1996. *Russia and the Idea of Europe.* London: Routledge.

Neumann, I.B. 1998. Identity and the outbreak of war. *International Journal of Peace Studies*, 3(1), 7–22.

Neumann, I.B. 2010. National security, culture and identity. In: V. Mauer and M.D. Cavelty (eds.) *The Routledge Handbook of Security Studies*, edited by. New York: Routledge, 95–104.

Neumann, I.B. 2011. Governing a great power: Russia's oddness reconsidered. In: I. Neumann and O.J. Sending (eds.) *Governing the Global Polity: Practice, mentality, rationality*. Ann Arbor, MI: University of Michigan Press, 70–110.

Nivat, A. 2001. *Chienne de Guerre. A woman reporter behind the lines of the war in Chechnya*. New York: PublicAffairs.

Omelicheva, M. 2009. Russia's counterterrorism legislation, warts and all. In: A. Mikkonen (ed.) *Threats and Prospects in Combating Terrorism: Series 2, Research Reports No 41*. Helsinki: National Defence University, Department of Strategic and Defence Studies, 1–20.

OSCE Representative on freedom of the media (ed.) 2003. *Freedom and Responsibility. Yearbook 2002–2003*. Vienna: OSCE.

Pain, E. 2005. The Chechen War in the context of contemporary Russian politics. In: R. Sakwa (ed.) *Chechnya from Past to Future*. London: Anthem Press, 67–78.

Peoples, C. and Vaughan-Williams, N. 2010. *Critical Security Studies: An introduction*. London: Routledge.

Petersson, B. 2003. Combating uncertainty, combating the global. In: B. Petersson and E. Clark (eds.) *Identity Dynamics and the Construction of Boundaries*. Lund: Nordic Academic Press, 99–121.

Pram Gad, U. and Lund Petersen, K. 2011. Concepts of politics in securitization theory. *Security Dialogue*, 42(4/5), 315–329.

Presidential Address to the Federal Assembly 3 December 2015, available at Kremlin.ru (http://en.kremlin.ru/events/president/news/50864).

Ram, H. 1999. *Prisoners of the Caucasus: Literary myths and media representations of the Chechen conflict*. Berkeley Program in Soviet and Post-Soviet Studies, Working Paper Series. Berkeley, CA: University of California.

Ringmar, E. 1996. *Identity, Interest and Action. A cultural explanation of Sweden's intervention in the Thirty Years War*. Cambridge: Cambridge University Press.

Roe, P. 2008. Actor, audiences and emergency measures: Securitization and the UK's decision to invade Iraq. *Security Dialogue*, 39(6), 615–635.

Russell, J. 2002. Mujahedeen, mafia, madmen … Russian perceptions of Chechens during the wars in Chechnya, 1994–1996 and 1999– to date. *Journal of Communist Studies and Transition Politics*, 18(1), 73–96.

Russell, J. 2005. Terrorists, bandits, spooks and thieves: Russian Demonisation of the Chechens prior to and since 9/11. *Third World Quarterly*, 26(1), 101–116.

Russell, J. 2007. *Chechnya – Russia's 'War on Terror'*. London: Routledge.

Sakwa, R. 2005. *Chechnya from Past to Future*. London: Anthem Press.

Salter, M.B. 2002. *Barbarians and Civilization in International Relations*. London: Pluto Press.

Salter, M.B. 2008. Securitization and desecuritization: Dramaturgical analysis and the Canadian Aviation Transport Security Authority. *Journal of International Relations and Development*, 11(4), 321–349.

Saunders, R. 2008. A conjurer's game: Vladimir Putin and the politics of presidential prestidigitation. In: G. Kassimeris (ed.) *Playing Politics with Terrorism: A user's guide*. New York: Columbia University Press, 220–249.

Saussure, F.D. 1974. *Course in General Linguistics*. London: Fontana.

Seely, R. 2001. *Russo-Chechen Conflict, 1800–2000*. London: Frank Cass.

Smith, G. 1999. The masks of Proteus: Russia, geopolitical shift and the new Eurasianism. *Transactions of the Institute of British Geographers, New Series*, 24(4), 481–494.

Smith, S. 1998. *Allah's Mountains. The battle for Chechnya*. London: Tauris Parke.

Snetkov, A. 2007. The image of the terrorist threat in the official Russian press: The Moscow Theatre crisis (2002) and the Beslan hostage crisis (2004). *Europe–Asia Studies*, 59(8), 1349–1365.

Snetkov, A. 2015. *Russia's Security Policy under Putin: A critical perspective*. London: Routledge.

Stephens, A.C. and Vaughan-Williams, N. (eds.) 2009. *Terrorism and the Politics of Response*. New York: Routledge.

Stritzel, H. 2007. Towards a theory of securitization: Copenhagen and beyond. *European Journal of International Relations*, 13(3), 357–383.

Stritzel, H. 2014. *Security in Translation: Securitization theory and the localization of threat*. London: Palgrave Macmillan.

Taureck, R. 2006. Securitization theory and securitization studies. *Journal of International Relations and Development*, 9(1), 52–61.

Thorup, M. 2010. *An Intellectual History of Terror: War, violence and the state*. London: Routledge.

Tishkov, V. 1997. *Ethnicity, Nationalism and Conflict in and after the Soviet Union: The mind aflame*. London: SAGE.

Tishkov, V. 2004. *Chechnya: Life in a war-torn society*. Berkeley, CA: University of California Press.

Toft, M.D. 2006. Issue indivisibility and time horizons as rationalist explanations of war. *Security Studies*, 15(1), 34–69.

Tolz, V. 1998. Forging the nation: National identity and national building in post-Communist Russia. *Europe–Asia Studies*, 50(6), 993–1022.

Torfing, J. 1999. *New Theories of Discourse: Laclau, Mouffe and Zizek*. Oxford: Blackwell.

Treisman, D. 1997. Russia's 'ethnic revival': The separatist activism of regional leaders in a postcommunist order. *World Politics*, 49(2), 212–249.

Trenin, D. and Malashenko, A. 2004. *Russia's Restless Frontier: The Chechnya factor in post-Soviet Russia*. Washington, DC: Carnegie Endowment for International Peace.

Tsygankov, A.P. 1997. From international institutionalism to revolutionary expansionism: The foreign policy discourse of contemporary Russia. *Mershon International Studies Review*, 41(2), 247–268.

Urban, J.B. and Solovei, V.D. 1997. *Russia's Communists at the Crossroads*. Boulder, CO: Westview Press.

Vuori, J. 2008. Illocutionary logic and strands of securitization: Applying the theory of securitization to the study of non-democratic political orders. *European Journal of International Relations*, 14(1), 65–99.

Wagnsson, C. 2000. *Russian Political Language and Public Opinion on the West, NATO and Chechnya*. Stockholm: Akademitryck AB Edsbruk.

Walker, R. 1990. Security, sovereignty and the challenge of world politics, *Alternatives*, 15(1), Spring 1990, 3–27.

Ware, R.B. and Kisriev, E. 2010. *Dagestan: Russian hegemomy and Islamic resistance in the North Caucasus*. New York: M.E. Sharp.

Weber, C. 2006. An aesthetics of fear: The 7/7 London bombings. *Millennium*, 34(3), 683–710.

White, S. and McAllister, I. 2006. Politics and the media in post-Communist Russia. In: K. Voltmer (ed.) *Mass Media and Political Communication in New Democracies*. New York: Routledge, 210–227.

Wilhelmsen, J. 1999. *Conflict in the Russian Federation: Two case studies, one Hobbesian explanation*. NUPI Report 249. Oslo: NUPI.

Wilhelmsen, J. 2003. *Norms: The forgotten factor in Russian–Western rapprochement. a case study of freedom of the press under Putin*. FFI Report 00457. Kjeller: FFI.

Wilhelmsen, J. 2005. Between a rock and a hard place: The Islamisation of the Chechen separatist movement. *Europe–Asia Studies*, 57(1), 35–59.

Wilhelmsen, J. 2011. Russia and international terrorism: Global challenge – national response. In: J. Wilhelmsen and E.W. Rowe (eds.) *Russia's Encounter with Globalisation*. Houndmills: Palgrave Macmillan, 97–134.

Wilhelmsen, J. 2016, How does war become a legitimate undertaking? Re-engaging the post-structuralist foundation of securitization theory. *Cooperation and Conflict*, Published online before print, May 12, 2016, doi: 10.1177/0010836716648725.

Wilkinson, C. 2007. The Copenhagen School on tour in Kyrgyzstan. *Security Dialogue*, 38(1), 5–25.

Williams, M.C. 2003. Words, images, enemies: Securitization and international politics. *International Studies Quarterly*, 47(4), 511–531.

Wæver, O., Busan, B., Kelstrup, M. and Lemaitre, P. (eds.) 1993. *Identity, Migration and the New Security Agenda in Europe*. London: Pinter.

Wæver, O. 1995a. Securitization and desecuritization. In: R.D. Lipschutz (ed.) *On Security*. New York: Columbia University Press, 46–86.

Wæver, O. 1995b. Identity, integration and security: Solving the sovereignty puzzle in EU studies. *Journal of International Affairs*, 48(2), 289–431.

Wæver, O. 1996. European security identities. *Journal of Common Market Studies*, 34(1), 103–132.

Wæver, O. 2002. Identity, communities and foreign policy: Discourse analysis as foreign policy theory. In: L. Hansen and O. Wæver (eds.) *European Integration and National Identity*. London: Routledge, 20–49.

Wæver, O. 2003. from 'Securitization: taking stock of a research programme in Security Studies'. Unpublished manuscript.

Wæver, O. 2006. What's religion got to do with it? Terrorism, war on terror, and global security. Keynote lecture at the *Nordic Conference on the Sociology of Religion*. Aarhus, 11 August 2006.

Zyuganov, G.A. 1992. *Drama Vlasti*. Moscow.

Index

Page numbers in *italics* denote tables, those in **bold** denote figures.

Taylor & Francis eBooks

Helping you to choose the right eBooks for your Library

Add Routledge titles to your library's digital collection today. Taylor and Francis ebooks contains over 50,000 titles in the Humanities, Social Sciences, Behavioural Sciences, Built Environment and Law.

Choose from a range of subject packages or create your own!

Benefits for you

» Free MARC records
» COUNTER-compliant usage statistics
» Flexible purchase and pricing options
» All titles DRM-free.

Benefits for your user

» Off-site, anytime access via Athens or referring URL
» Print or copy pages or chapters
» Full content search
» Bookmark, highlight and annotate text
» Access to thousands of pages of quality research at the click of a button.

> **REQUEST YOUR FREE INSTITUTIONAL TRIAL TODAY**
>
> **Free Trials Available**
> We offer free trials to qualifying academic, corporate and government customers.

eCollections – Choose from over 30 subject eCollections, including:

Archaeology	Language Learning
Architecture	Law
Asian Studies	Literature
Business & Management	Media & Communication
Classical Studies	Middle East Studies
Construction	Music
Creative & Media Arts	Philosophy
Criminology & Criminal Justice	Planning
Economics	Politics
Education	Psychology & Mental Health
Energy	Religion
Engineering	Security
English Language & Linguistics	Social Work
Environment & Sustainability	Sociology
Geography	Sport
Health Studies	Theatre & Performance
History	Tourism, Hospitality & Events

For more information, pricing enquiries or to order a free trial, please contact your local sales team:
www.tandfebooks.com/page/sales

 Routledge
Taylor & Francis Group

The home of
Routledge books

www.tandfebooks.com